Probability Theory - A Comprehensive Course

Probability Theory - A Comprehensive Course

Editor

Giorgos Michel

scitus
academics

Probability Theory - A Comprehensive Course
Edited by **Giorgos Michel**

ISBN: 978-1-68117-453-2
Library of Congress Control Number: 2016937612

SCITUS Academics LLC,
www.scitusacademics.com
Box No. 4766, 616 Corporate Way,
Suite 2, Valley Cottage,
NY 10989

Printed in United States of America on Acid Free Paper

Preface

Probability theory is the branch of mathematics concerned with probability, the analysis of random phenomena. The central objects of probability theory are random variables, stochastic processes, and events: mathematical abstractions of non-deterministic events or measured quantities that may either be single occurrences or evolve over time in an apparently random fashion. The outcome of a random event cannot be determined before it occurs, but it may be any one of several possible outcomes. The actual outcome is considered to be determined by chance. The word probability has several meanings in ordinary conversation. Two of these are particularly important for the development and applications of the mathematical theory of probability. One is the interpretation of probabilities as relative frequencies, for which simple games involving coins, cards, dice, and roulette wheels provide examples. It is not possible to predict precisely results of random event. However, if a sequence of individual events, such as coin flipping or the roll of dice, is influenced by other factors, such as friction, it will exhibit certain patterns, which can be studied and predicted. Two representative mathematical results describing such patterns are the law of large numbers and the central limit theorem. As a mathematical foundation for statistics, probability theory is essential to many human activities that involve quantitative analysis of large sets of data. Methods of probability theory also apply to descriptions of complex systems given only partial knowledge of their state, as in statistical mechanics. The book Probability Theory: A Comprehensive Course emphasizes modern probability theory. Overall, probabilistic concepts play an increasingly important role in mathematics, physics, biology, financial engineering and computer science. They help us in understanding magnetism, amorphous media, genetic diversity and the perils of random developments at financial markets, and they guide us in constructing more efficient algorithms.

Table of Contents

CHAPTER 1

Born Rule and Noncontextual Probability

Fabrizio Logiurato[1,2*], Augusto Smerzi[2]

[1]INO-CNR BEC Center and Physics Department, Trento University, Povo, Italy

[2]QSTAR, INO-CNR and LENS, Largo Enrico Fermi, Firenze, Italy

ABSTRACT

We present a new derivation of the Born rule from the assumption of noncontextual probability (NCP). Within the theorem we also demonstrate the continuity of probability with respect to the amplitudes, which has been suggested to be a gap in Zurek's and Deutsch's approaches, and we show that NCP is implicitly postulated also in their derivations. Finally, physical motivations of NCP are given based on an invariance principle with respect to a resolution change of measurements and with respect to the principle of no-faster-than-light signalling.

KEYWORDS

Born Rule; Non Contextual Probability

1. INTRODUCTION

The fundamental probabilistic postulate which allows QM its successful predictions was first introduced by Born (from a suggestion by Einstein) in 1926 within the scattering theory [1]. The question if such postulate could be derived as a theorem from other postulates of QM was quickly posed. A first answer was provided by Von Neumann and included in his famous book of 1932 on the mathematical foundations of QM [2]. His theorem nowadays does not enjoy much attention since it is believed to contain more assumptions than Gleason's [3,4]. Different attempts have been discussed in the context of the relative-state [5] and the many-worlds interpretations [6] of QM by Finkelstein [7] and Hartle [8], which have analyzed an endless sequence of measurements to show that the relative frequency follows the Born rule. However, the meaning in the real world of an infinite sequence of measurements is controversial and it is doubtful whether these proofs contain circularities [9-14].

The 1957 Gleason work [4] is regarded as the most important derivation of the Born rule [15], but this theorem is usually considered quite formal and difficult to grasp. Moreover, it is based on the concept of noncontextual probability (NCP), whose connection with physics is not clear. In the structure of the Gleason theorem, every Hilbert subspace $\mathcal{M}$ corresponds to an observable quantity. Each one of such subspaces is representable by a projection operator $\hat{\Pi}_{\mathcal{M}}$. A projector $\hat{\Pi} = |\psi\rangle\langle\psi|$ can represent the question with answer yes or not concerning an experiment testing whether the system has the respective state $|\psi\rangle$. A complete set of mutually commuting projectors $\{\hat{\Pi}_i\}$ corresponds to a set of questions which can simultaneously be asked in a measurement. In the logical-algebraic approach [16], probability is a measure defined over a projection lattice (set of closed subspaces) of Hilbert space $\mathcal{H}$, a mapping $p:\hat{\Pi}_M \to [0,1]$. The postulates on which the theorem rests are:

1) The probability assigned to a complete set of N projectors $\{\hat{\Pi}_i\}$ is normalized:

$$p(I)=1,\ \sum_{i=1}^{N}\hat{\Pi}_i = I. \tag{1}$$

2) For any sequence of M mutually orthogonal projectors:

$$p\left(\sum_{i=1}^{M}\hat{\Pi}_i\right)=\sum_{i=1}^{M}p\left(\hat{\Pi}_i\right). \tag{2}$$

Assuming Postulate 2) is equivalent to postulating NCP for projective measurements (or, i.e., according to the definition by Spekkens, measurement noncontextuality [17]): the probability of a certain occurrence, for example the answer yes for a projector, does not depend on other questions simultaneously tested, that is, on other projectors. In terms of basis vectors, the probability of obtaining a given state is independent of the basis it belongs to. To illustrate this important point, let us consider, for instance, two complete set of mutually orthogonal projectors $\{\hat{\Pi}_i\}$, $\{\hat{\Pi}'_i\}$ with $N=3$, where $\hat{\Pi}_1 \equiv \hat{\Pi}'_1$. From the normalization postulate of probability:

$$p\left(\hat{\Pi}_1\right)+p\left(\hat{\Pi}_2\right)+p\left(\hat{\Pi}_3\right)=1, \tag{3}$$

$$p'(\hat{\Pi}_1) + p(\hat{\Pi}'_2) + p(\hat{\Pi}'_3) = 1, \tag{4}$$

where in principle $p(\hat{\Pi}_1)$ and $p'(\hat{\Pi}_1)$ may be different. In the subspace orthogonal to $\hat{\Pi}_1$ we have

$\hat{\Pi}_2 + \hat{\Pi}_3 = \hat{\Pi}'_2 + \hat{\Pi}'_3$, then $p(\hat{\Pi}_2 + \hat{\Pi}_3) = p(\hat{\Pi}'_2 + \hat{\Pi}'_3)$

and from Postulate 2:

$$p(\hat{\Pi}_2) + p(\hat{\Pi}_3) = p(\hat{\Pi}'_2) + p(\hat{\Pi}'_3). \tag{5}$$

Then, by comparing Equation (3) and Equation (4), we get $p(\hat{\Pi}_1) = p'(\hat{\Pi}_1)$. It is also possible to show the vice versa, i.e, if NCP holds then Postulate 2 is true.

From Postulates 1 and 2 it follows Gleason's Theorem: For a Hilbert space with dimension $\dim\mathcal{H} \geq 3$ there exists a density operator $\hat{\rho}$ which acts on $\mathcal{H}$ such that the probability measure for every projector $\hat{\Pi}$ is the trace rule:

$$p(\hat{\Pi}) = tr(\hat{\rho}\hat{\Pi}), \tag{6}$$

namely, the Born rule.

More recently, new proofs appeared in the literature with the intention of being "more physically motivated than the theorem of Gleason" [18]. Their aim, as the ours in the present work, is to shed new light on the physical principles of QM, particularly on the origin of the Born rule. Indeed, Gleason's theorem gives "rather little insight into the emergence of quantum probabilities and the Born rule" [19].

Particularly interesting are Deutsch's and Zurek's derivations [18,20-25]. Their approaches to the Born rule have similar structures. As a first step, they use an invariance principle in order to show that equal amplitudes correspond to equal probabilities. Deutsch finds such principle in the theory of decisions (for critical discussions and further elaborations see [26-28]). Zurek introduces an environment-assisted invariance principle [29,30], envariance, in the framework of a relative-state approach where the system $\mathcal{S}$ under measurement jointly evolves with the environment $\mathcal{E}$ and, possibly, with one or more auxiliary measurement devices. After demonstrating the correspondence between equal amplitudes and equal probabilities, both Zurek and Deutsch consider a finegraining technique to deal with the general case of different amplitudes of the initial state. For this purpose they have to introduce auxiliary systems which become entangled with $\mathcal{S}$.

However, fine-graining has been questioned: Caves [23] considers disturbing the necessity of introducing additional systems of adequate dimensions, possibly in an infinite Hilbert space, in order to reach the wanted approximation to irrational probabilities. Barnum [25] points out an even more stringent weakness: the step from rational to irrational amplitudes requires the continuity of the probability with respect to such amplitudes and such a property is not demonstrated by Zurek and Deutsch. We note that a fundamental and extensive part of Gleason's theorem is actually devoted to prove that the probability is continuous. For several other controversial points of Zurek's demonstration, the most of them shares with Deutsch's, we refer to [19,23-25].

So, although Gleason-type theorems are usually considered well established, a simple, physically justified and shared derivation of the Born rule is not yet present in the literature. With our work we intend to fill this gap.

In the next section we propose a new derivation of the Born rule from the NCP assumption for projective measurements and pure states, the generalization to mixed states being immediate. Whereas Gleason assumes that a density operator acting on $\mathcal{H}$ describes a general quantum state of the system only at the final step of his work, we introduce vectors of the Hilbert space describing quantum states from the very beginning of our derivation. On the one hand it allows a much more elementary deduction of the Born rule, and on the other hand it makes the postulates and the physical principles of our theorem easier to analyze in comparison with the previous demonstrations. Moreover, involving Zurek's and Deutsch's demonstrations pure states, also the connections between those works, the Gleason Theorem and the ours are clearer.

We do not enter into the controversy about the origin and the meaning of probability, and following Gleason and others (see also Mohrhoff in [24]), we assume the existence of probability defined on a quantum state. In addition, differently by Zurek, we do not suppose unitary evolution and consider a single system. As Gleason's Theorem, our result holds for a Hilbert space with $\dim\mathcal{H} \geq 3$, whereas in general, because of fine-graining, Zurek's and Deutsch's hold for $\dim\mathcal{H} = \infty$. We also give a rather simple derivation that the probability is continuous from NCP. Without this proof, Deutch's and Zurek's results only hold for rational values of amplitudes.

In Section III we generalize our theorem to include entangled states and multi-particles systems. Since the Hilbert space $\mathcal{H}$ of such systems has dimension which is the product of the dimensions of its components, the coupling of a subsystem with $\dim\mathcal{H} \geq 3$ with another subsystem with $\dim\mathcal{H} = 2$ enables us to extend the Born rule also to such space.

In Section IV we show that the Gleason postulate of NCP occurs in the use of fine-graining and that such a postulate is a hidden assumption of Zurek's and Deutsch's proofs. Indeed, fine-graining is a resolution change of our measurements and NCP is equivalent to a condition of invariance with respect to such a change: finegraining on a subspace of the state must not alter the probability of finding the system in a different subspace. This resolution invariance is an intuitive and physically natural interpretation of measurement noncontextuality.

Finally, in Section V, we show that the condition of noncontextual probability, or resolution invariance, can also be deduced by the hypothesis of no-faster-than-light signalling of the Special Relativity.

2. BORN RULE AS A THEOREM FROM NONCONTEXTUAL PROBABILITY

Contrarily to the principles of classical mechanics, the axioms of QM do not have an equally shared and clear formulation. We do not want to enter into such a thorny issue, related to the interpretative problems of QM. For our purpose, from the quantum postulates generally considered in standard presentations of QM (see, e.g., [31-34]), we can extract the following common assumptions:

Postulate I:

At the time t the state of every physical system $\mathcal{S}$ is described by a normalized vector $|\psi_S(t)\rangle$ belonging to a Hilbert space $\mathcal{H}$.

Postulate II:

Every measurable physical quantity $\mathcal{A}$ is described by a Hermitian operator $\hat{A}$ acting in $\mathcal{H}$.

Postulate III:

The only possible result of a measurement of an observable $\mathcal{A}$ is one of the eigenvalues $\{a_i\}$ of $\hat{A}$, (as the customary habit, from now on we will identify the observable with its operator).

Postulate IV:

If the measurement of the observable $\hat{A}$ on the system in the state $|\psi_S(t)\rangle$ gives the result a_i, the state of the system immediately after the measurement is the associate normalized eigenstate $|a_i\rangle$ (non-degenerate case).

Postulate I is known as the Completeness Postulate of QM. Postulate II links an observable quantity with an operator in a Hilbert space (observable-operator link). Postulate III connects a particular value of that observable with an eigenvalue of the corresponding operator (value-eigenvalue link). Postulate IV relates a detected eigenvalue with the state of the system (eigenvalueeigenstate link).

Now consider a Hermitian observable $\hat{A}$ with discrete eigenvalues $\{a_i\}$ and eigenstates $\{|a_i\rangle\}$, $i = 1, \cdots, N$, and a single system $\mathcal{S}$ described by the state $|\psi_S\rangle \in \mathcal{H}^N$. $\{|a_i\rangle\}$ is a complete orthonormal set and $|\psi_S\rangle$ can be decomposed into components:

$$|\psi_S\rangle = \sum_{i=1}^{N} c_i |a_i\rangle. \tag{7}$$

We intend to prove the following

Theorem 1:

If NCP and Postulates I-IV hold, then for $N \geq 3$ the probability of obtaining the non-degenerate eigenvalue a_i, measuring $\hat{A}$ on the state $|\psi_S\rangle$, is necessarily given by the square modulus of the inner product:

$$p(a_i) = |\langle a_i|\psi_S(t)\rangle|^2 \quad \text{Born Rule,} \tag{8}$$

where $|a_i\rangle$ is the eigenstate associated with the eigenvalue a_i.

We assume the probability as a primitive concept and we begin the derivation of Theorem 1 by proving some simple lemmas. We underline that any theorem having the aim to prove the Born rule has to face the fact that the connection between probability and quantum state must be postulated or deduced by other means. Gleason, for instance, postulates the probability to be a measure defined on states of the basis. As Gleason, we suppose QM to be a probabilistic theory, which makes predictions about probabilities of occurrences of observable quantities, and that probabilities are functions of quantum states. Then, from QM postulates we want to show that

Lemma 1:

The probability of obtaining an eigenvalue a_i is equal to the probability of obtaining the respective eigenstate $|a_i\rangle$ (statistical case):

$$p(a_i) = p(|a_i\rangle). \tag{9}$$

wang#title3_4:spProof:

Suppose that the system is prepared with the eigenvalue a_i, in this case $p(a_i)=1$. According to Postulate IV if $p(a_i)=1$ then $p(|a_i\rangle)=1$, and vice versa. But $\{a_i\}$ correspond to mutually exclusive events, thus has to be $p(a_j)=0$ for $j \neq i$. Then must necessarily be $p(|a_j\rangle)=0$ too, since, otherwise, having obtained with a measurement the state $|a_j\rangle$ we should assign to the system the eigenvalue a_j, in contradiction with the exclusivity property of $\{a_i\}$ events. Therefore we have the equations for the conditional probabilities for a non-statistical case:

$$p(|a_i\rangle;a_j) = p(a_j;|a_i\rangle) = \delta_{ij}, \tag{10}$$

which represent the probability of finding the state $|a_i\rangle$ for systems prepared with certainty with eigenvalue a_j, and vice versa. Because of Postulate III, the only possible result of a measurement is one of the eigenvalues of $\hat{A}$. Consider then a generic preparation $|\psi_S\rangle$ of the initial state with $p(|a_i\rangle) \neq 1$. Introducing the joint probability $p(|a_i\rangle,a_j)$ and considering the general validity of the Bayesian formula for the joint probabilities $p(|a_i\rangle,a_j) = p(|a_i\rangle;a_j)\,p(a_j)$, we have

$$\begin{aligned} p(|a_i\rangle) &= \sum_{j=1}^{N} p(|a_i\rangle,a_j) \\ &= \sum_{j=1}^{N} p(|a_i\rangle;a_j)\,p(a_j) = p(a_i), \end{aligned} \tag{11}$$

where we have used Equation (10) in the conditional probabilities of Equation (11).

We note from Equations (10) and (11) that

$p(|a_i\rangle) = p(|a_i\rangle, a_i)$ and that $\{|a_i\rangle\}$ represent a set of complete and mutually exclusive events:

$$p(|a_i\rangle, |a_j\rangle) = \delta_{ij}, \qquad \sum_{i=1}^{N} p(|a_i\rangle) = 1. \tag{12}$$

Lemma 1 can be easily generalized for entangled states as well. For instance, consider the state:

$$|\psi_{SB}\rangle = \sum_{i=1}^{N} \sum_{k=1}^{M} c_{ik} |a_i\rangle |b_k\rangle, \tag{13}$$

where $\hat{B}$ is a new observable commuting with $\hat{A}$, with eigenvalues $\{b_k\}$ and eigenstates $\{|b_k\rangle\}$, $k = 1, \cdots, M$, and $|\psi_{SB}\rangle \in \mathcal{H}^{N} \otimes \mathcal{H}^{M}$, (we point out that $\hat{B}$ does not necessarily have to belong to another system: for example, $\hat{A}$ could be the spin observable and $\hat{B}$ the position observable of one particle). Consider the pairs of eigenvalues $\{a_i, b_k\}$, and eigenstates $\{|a_i\rangle, |b_k\rangle\}$. By using conditional probabilities, it follows that on the generic state $|\psi_{SB}\rangle$:

$$\begin{aligned} p(|a_i\rangle |b_k\rangle) &= p(|a_i\rangle, |b_k\rangle) \\ &= p(\{a_i, b_k\}) \end{aligned} \tag{14}$$

(see Appendix A for a proof of this). Then the probability $p(a_i)$ is again given by:

$$p(a_i) = \sum_{k=1}^{M} p(\{a_i, b_k\}) = \sum_{k=1}^{M} p(|a_i\rangle |b_k\rangle)$$
$$= \sum_{k=1}^{M} p(\{|a_i\rangle, |b_k\rangle\}) = p(|a_i\rangle). \quad (15)$$

With a similar reasoning we can also show that $p(b_k) = p(|b_k\rangle)$. We notice from Equations (14) and (15) that the outcomes of $\{|a_i\rangle, |b_k\rangle\}$ are mutually exclusive events as well, with the condition of normalization following from the normalization of $p(a_i)$.

Lemma 2: If a coefficient of an eigenstate in Equation (7) is zero, the probability of obtaining the corresponding eigenvalue and eigenstate is zero as well.

Proof: If, for instance, $c_k = 0$, the corresponding state $|a_k\rangle$ will be orthogonal to $|\psi_S\rangle$. We can think $\{|\psi_S\rangle, |a_k\rangle\}$ as eigenstates of the projector $\hat{\Pi}_S = |\psi_S\rangle\langle\psi_S|$ with eigenvalues $\{1,0\}$. The Hermitian operator $\hat{\Pi}_S$ represents the observable testing whether the system $\mathcal{S}$ is prepared with the state $|\psi_S\rangle$ or not.

The pair $\{|\psi_S\rangle, |a_k\rangle\}$ is a complete orthonormal basis of a subspace $\mathcal{H}^2$ of $\mathcal{H}^N$.

With the system prepared in the state $|\psi_S\rangle$:

$p(|\psi_S\rangle; |\psi_S\rangle) = 1$. From Lemma 1, it follows that

$p(1;|\psi_S\rangle)=1$ and because eigenvalues are exclusive events $p(0;|\psi_S\rangle)=0$. So, if we apply Lemma 1 again to the previous equation, considering that $|a_k\rangle$ is eigenstate of both $\hat{\Pi}_S$ and $\hat{A}$, we have $p(|a_k\rangle;|\psi_S\rangle)=0$ and $p(a_k;|\psi_S\rangle)=0$.

Also Lemma 2 can be directly generalized to entangled states. For instance, if in the state $|\psi_{SB}\rangle$ the coefficient $c_{kl}=0$ then $p(|a_k\rangle|b_l\rangle;|\psi_{SB}\rangle)=0$. In particular, if the state of the system is in a Schmidt decomposition:

$$|\psi_{SB}\rangle=\sum_{i=1}^{N}c_i|a_i\rangle|b_i\rangle, \tag{16}$$

from Lemma 1 and Lemma 2, since $p(|a_i\rangle|b_k\rangle)=0$ for $k\neq i$, it follows the perfect correlation property $p(|a_i\rangle)=p(|b_i\rangle)$:

$$\begin{aligned} p(a_i)&=\sum_{k=1}^{M}p(\{a_i,b_k\})=p(|a_i\rangle|b_i\rangle)=p(|a_i\rangle) \\ &=p(|b_i\rangle)=p(b_i). \end{aligned} \tag{17}$$

Lemma 3:

If the measurement is noncontextual then the probability of obtaining the eigenvalue a_i can only depend on the modulus of the coefficient c_i:

$$p(a_i)=p(|c_i|). \tag{18}$$

wang#title3_4:spProof:

Let us recall the meaning of NCP with an example (we refer to the work by Spekkens for an extensive analysis of the contextuality concept in arbitrary operational theories and in QM [17]). Consider two operators $\hat{A}$ and $\hat{A}'$ which do not commute, $[\hat{A},\hat{A}'] \neq 0$, in a Hilbert space with $\dim\mathcal{H} = 3$. Eigenvalues and eigenstates of $\hat{A}$ and $\hat{A}'$ are, respectively:

$$\begin{aligned} \hat{A}: &\ \{a_1, a_2, a_3\}, \ \ \{|a_1\rangle, |a_2\rangle, |a_3\rangle\}, \\ \hat{A}': &\ \{a_1, a_2', a_3'\}, \ \ \{|a_1\rangle, |a_2'\rangle, |a_3'\rangle\}, \end{aligned} \tag{19}$$

where they have in common the eigenvalue a_1 and the associate eigenstate (we see that this is possible, if $\hat{A}$ and $\hat{A}'$ must not commute, just for $N \geq 3$). We can write the state $|\psi_S\rangle$ in the two different bases of Equation (19):

$$\begin{aligned} |\psi_S\rangle &= c_1|a_1\rangle + c_2|a_2\rangle + c_3|a_3\rangle \\ &= c_1|a_1\rangle + c_2'|a_2'\rangle + c_3'|a_3'\rangle. \end{aligned} \tag{20}$$

In general, the probability of obtaining the eigenvalue a_1 could depend on whether we measure the observable $\hat{A}$ or $\hat{A}'$. In other words, the probability $p(a_1)$ could be different if we were in the experimental context which is prepared for measuring which value $\hat{A}$ assumes of the triplet $\{a_1, a_2, a_3\}$ or which value $\hat{A}'$ assumes of the triple $\{a_1, a_2', a_3'\}$.

With the result of Lemma 1 and the completeness postulate of QM, we can write the more general probability in different experimental contexts as:

$$p_A(a_1) = p\left(|a_1\rangle; |\psi_S\rangle, \{|a_1\rangle, |a_2\rangle, |a_3\rangle\}\right),$$
$$p_{A'}(a_1) = p\left(|a_1\rangle; |\psi_S\rangle, \{|a_1\rangle, |a_2'\rangle, |a_3'\rangle\}\right). \tag{21}$$

The probability is noncontextual, and we have measurement noncontextuality, if $p(a_1)$ does not depend on the measurement with other eigenvalues:

$$p_A(a_1) \equiv p_{A'}(a_1), \tag{22}$$

or, according to Lemma 1, the probability $p(|a_1\rangle)$ does not depend on the measurement with other eigenstates:

$$p_A(|a_1\rangle) \equiv p_{A'}(|a_1\rangle). \tag{23}$$

Then, by assuming NCP, Equations (21) can be reduced to the equation:

$$p(a_1) = p\left(|a_1\rangle; |\psi_S\rangle\right), \tag{24}$$

stating that the probability only depends on the eigenstate $|a_1\rangle$ and, from the completeness postulate of QM and NCP, on the state of the system $|\psi_S\rangle$.

Thus, let us choose a new basis $\{|a_i'\rangle\}$ such that the state in Equation (7) is written with only two components:

$$|\psi_S\rangle = c_1|a_1\rangle + |c_2'||a_2'\rangle, \tag{25}$$

where $c_i' = 0$ for $i \geq 3$, $|c_2'|$ is the modulus of the amplitude c_2' and

$$|a_1'\rangle \equiv |a_1\rangle, \quad |a_2'\rangle = \frac{1}{|c_2'|}\sum_{i=2}^{N} c_i|a_i\rangle. \tag{26}$$

$\{|a_1\rangle, |a_2'\rangle\}$ could be regarded as eigenstates of an observable $\hat{A}' = |a_1\rangle\langle a_1|$ with eigenvalues $\{1,0\}$ checking whether our system is in the state $|a_1\rangle$ or not.

According to NCP the transformation of basis $\{|a_i\rangle\} \to \{|a_i'\rangle\}$ of the state cannot change the probability:

$$\begin{aligned} p(|a_1\rangle;|\psi_S\rangle) &= p(|a_1\rangle;|\psi_S\rangle)_{\{|a_i\rangle\}} \\ &= p(|a_1\rangle;|\psi_S\rangle)_{\{|a_i'\rangle\}}. \end{aligned} \tag{27}$$

Now we expressly write the phase of the coefficient c_1 in the state of Equation (25):

$$|\psi_S\rangle = \mathrm{e}^{\phi_1}|c_1||a_1\rangle + |c_2'||a_2'\rangle. \tag{28}$$

We get for the state $|a_1\rangle$ the probability:

$$p(|a_1\rangle;|\psi_S\rangle) = p(|a_1\rangle;\mathrm{e}^{\phi_1}|c_1||a_1\rangle + |c_2'||a_2'\rangle). \tag{29}$$

Therefore, considering the normalization $|c_1|^2 + |c_2'|^2 = 1$ and that the components of the orthonormal basis are the constants $\langle a_i' | a_j' \rangle = \delta_{ij}$, the most general function of the probability in Equation (29) can be written as:

$$p(|a_1\rangle; |\psi_S\rangle) = p(\phi_1, |c_1|). \quad (30)$$

A similar argument when applied to every state $|a_i\rangle$ gives

$$p(|a_i\rangle; |\psi_S\rangle) = p(\phi_i, |c_i|). \quad (31)$$

However, by using the conservation of probability we can show the independence from the phase of $p(|a_i\rangle; |\psi_S\rangle)$. Indeed, let us suppose to have two different states $|\psi_S\rangle$ and $|\psi_S^*\rangle$ which are distinct just with respect to the phase of the component $|a_1\rangle$ (to simplify the notation, we select a basis such that the components of the states are different from zero only in a subspace with $\dim\mathcal{H} = 2$):

$$\begin{aligned} |\psi_S\rangle &= e^{\phi_1} |c_1| |a_1\rangle + e^{\phi_2} |c_2| |a_2\rangle, \\ |\psi_S^*\rangle &= e^{\phi_1^*} |c_1| |a_1\rangle + e^{\phi_2} |c_2| |a_2\rangle, \end{aligned} \quad (32)$$

where with ϕ_1^* we have denoted the new phase. The conservation of the probability requires:

$$\begin{aligned} p(|a_1\rangle; |\psi_S\rangle) + p(|a_2\rangle; |\psi_S\rangle) &= 1, \\ p(|a_1\rangle; |\psi_S^*\rangle) + p(|a_2\rangle; |\psi_S^*\rangle) &= 1, \end{aligned} \quad (33)$$

where from Lemma 2 we have used the fact that $p(|a_i\rangle;|\psi_S\rangle)=0$ if $|a_i\rangle \perp |\psi_S\rangle$. Thanks to Equation (31), we have that

$$p(|a_2\rangle;|\psi_S\rangle)=p(|a_2\rangle;|\psi_S^*\rangle). \tag{34}$$

From this and Equation (33), we also get:

$$p(|a_1\rangle;|\psi_S\rangle)=p(|a_1\rangle;|\psi_S^*\rangle), \tag{35}$$

which is equivalent to write:

$$p(\phi_1^*,|c_1|)=p(\phi_1,|c_1|). \tag{36}$$

Therefore, since ϕ_1^* is an arbitrary phase, the probability $p(a_1)$ does not depend on it but can only depend on the modulus $|c_1|$. So, we can set $p(a_1)=p(|c_1|)$, and this proves our lemma (the generalization to probabilities of other eigenvalues a_i is immediate).

Now we have all the ingredients to prove our

Theorem 1:

The probability $p(a_i)$ is given by the square modulus of coefficient c_i.

Proof:

Let us consider the state $|\psi_S\rangle \in \mathcal{H}^N$ with $N \geq 3$ in Equations (7). For the sake of simplicity we focus our attention again on $p(a_1)$. We choose two bases $\{|a_i\rangle\}$ and $\{|a_i'\rangle\}$ such that

$$|\psi_S\rangle = c_1|a_1\rangle + c_2|a_2\rangle + c_3|a_3\rangle \tag{37}$$

$$= c_1|a_1\rangle + c_2'|a_2'\rangle, \tag{38}$$

where $c_i = 0$ for $i \geq 4$, $c_i' = 0$ for $i \geq 3$ and $|a_1\rangle = |a_1'\rangle$. From the conservation of the total probability for the states of the two different bases we get respectively:

$$p(|a_1\rangle) + p(|a_2\rangle) + p(|a_3\rangle) = 1, \tag{39}$$

$$p(|a_1\rangle) + p(|a_2'\rangle) = 1, \tag{40}$$

where from NCP we have the same probability $p(|a_1\rangle)$ in both Equations (39) and (40). With a change of variable we can write Equation (18) as $p(a_i) = f(|c_i|^2)$. From Equations (39) and (40) it follows that

$$\begin{aligned} f(|c_1|^2) + f(|c_2|^2) + f(|c_3|^2) &= 1, \\ f(|c_1|^2) + f(|c_2'|^2) &= 1. \end{aligned} \tag{41}$$

By comparing the two Equations (41) we get:

$$f\left(|c_2'|^2\right)=f\left(|c_2|^2\right)+f\left(|c_3|^2\right). \tag{42}$$

Since the state $|\psi_S\rangle$ is normalized, we also have

$$|c_1|^2+|c_2|^2+|c_3|^2=1, \tag{43}$$

$$|c_1|^2+|c_2'|^2=1, \tag{44}$$

and

$$|c_2'|^2=|c_2|^2+|c_3|^2. \tag{45}$$

Putting this in Equation (42):

$$f\left(|c_2|^2+|c_3|^2\right)=f\left(|c_2|^2\right)+f\left(|c_3|^2\right). \tag{46}$$

Now we recall that a function $f(x)$ is linear with respect to the variable x if and only if the two following properties are jointly satisfied:

$$\begin{aligned} &f(x_1+x_2)=f(x_1)+f(x_2) \quad \text{additivity,} \\ &f(\alpha x)=\alpha f(x) \qquad\qquad\quad \text{homogeneity,} \end{aligned} \tag{47}$$

where α is any real number. We remark that if α is a rational number, it can be shown that the homogeneity follows from the additivity. Furthermore, since rational numbers form a subset dense in the set of real numbers, such derivation can be extended to the case of irrational α, if f is a continuous function [35].

Therefore, the condition of additivity is sufficient to establish the linearity if f is a continuous function. On the contrary, if f is not assumed to be continuous, the condition of additivity implies linearity only for rational values of the variable x.

Hence, if we assume that the probability is continuous with respect to the coefficients $|c_i|^2$, Equation (46) is a constraint of linearity for the probability. In this case we can write:

$$p(a_i) = k|c_i|^2, \quad i = 2,3, \tag{48}$$

where k denotes a constant. From Lemma 1 we have that $p(|c_i|^2 = 1) = 1$ and therefore $k = 1$ and $p(a_i) = |c_i|^2$. Hence, the Born rule is deduced for continuous probabilities.

We point out that we have postulated the continuity of probability. However, a derivation of such a property from noncontextuality is present in Gleason's work and it occupies a prominent part of that paper. Therefore in comparison to Gleason we have only accomplished half the job: as Bertrand Russell once said, postulating is equivalent to theft on the honest fatigue. On the other hand, if the continuity of probability is neither postulated nor demonstrated, we can only say that the Born rule applies only for rational values of the coefficients $|c_i|^2$.

Thus we now intend to complete our derivation by showing from NCP that the probability is continuous and that the Born rule also holds for irrational coefficients.

Consider, once again, $p(|a_1\rangle)$. We select a basis

$\{|a_i'\rangle\}$ such that the state $|\psi_S\rangle$ is written with two components as in Equation (38), with $|a_1'\rangle \equiv |a_1\rangle$, and the conservation of probability is given by Equation (40).

Then we choose a new basis $\{|a_i''\rangle\}$ in general with $|a_1''\rangle \neq |a_1\rangle$, $c_i'' = 0$ for $i \geq 4$, but with $|a_2''\rangle \equiv |a_2'\rangle$ so that we can write

$$|\psi_S\rangle = c_1''|a_1''\rangle + c_2'|a_2'\rangle + c_3''|a_3''\rangle. \tag{49}$$

The conservation of probability for the new states of the basis requires that

$$p(|a_1''\rangle) + p(|a_2'\rangle) + p(|a_3''\rangle) = 1. \tag{50}$$

From the previous condition we have the inequality:

$$p(|a_1''\rangle) + p(|a_2'\rangle) \leq 1. \tag{51}$$

By comparing Equation (40) with Equation (51) and since, thanks to NCP, the probability $p(|a_2'\rangle)$ is the same in these equations, we get

$$p(|a_1''\rangle) \leq p(|a_1\rangle). \tag{52}$$

Finally, we choose a third basis $\{|a_i'''\rangle\}$ such that $c_i''' = 0$ for $i \geq 4$, this time with $|a_1'''\rangle \equiv |a_1\rangle$:

$$|\psi_S\rangle = c_1|a_1\rangle + c_2'''|a_2'''\rangle + c_3'''|a_3'''\rangle. \tag{53}$$

The conservation of probability gives us:

$$p(|a_1\rangle) + p(|a_2'''\rangle) + p(|a_3'''\rangle) = 1. \tag{54}$$

From Equation (54) we get the inequality:

$$p(|a_1\rangle) \le 1 - p(|a_2'''\rangle). \tag{55}$$

Therefore the probability $p(|a_1\rangle)$ is in the range:

$$p(|a_1''\rangle) \le p(|a_1\rangle) \le 1 - p(|a_2'''\rangle). \tag{56}$$

Suppose $|c_1|^2$ to be an irrational number, we choose rational values for $|c_1''|^2$ and $|c_2'''|^2$ so that $p(|a_1''\rangle) = |c_1''|^2$ and $p(|a_2'''\rangle) = |c_2'''|^2$ By putting $p(|a_1\rangle) = p(|c_1|^2)$, Equation (56) becomes

$$|c_1''|^2 \le p(|c_1|^2) \le 1 - |c_2'''|^2, \quad |c_1''|^2, |c_2'''|^2 \in \mathbb{Q}, \tag{57}$$

with $|c_1''|^2 \le 1 - |c_2'''|^2$. Let us denote with $|\psi_{S\perp}\rangle$ the orthonormal state to $|\psi_S\rangle$. Consider the limits $|a_1''\rangle \to |a_1\rangle$, $|a_3''\rangle \to |\psi_{S\perp}\rangle$ and $|a_2'''\rangle \to |a_2'\rangle$, $|a_3'''\rangle \to |\psi_{S\perp}\rangle$. From Equation (38) we have $|c_1''|^2 \to |c_1|^2$, $|c_3''|^2 \to 0$ and $|c_2'''|^2 \to |c_2'|^2$, $|c_3'''|^2 \to 0$. Summarizing, with the normalization condition in Equation (44) we get:

$$|c_1''|^2 \to |c_1|^2, \quad 1 - |c_2'''|^2 \to |c_1|^2. \tag{58}$$

Because rational numbers form a dense set in the set of real numbers, we can always find a couple of numbers $|c_1''|^2, 1-|c_2'''|^2 \in \mathbb{Q}$, which are as near as we want to the irrational number $|c_1|^2$. Therefore, from Equation (57) it follows that the oscillation of the function p is zero at $|c_1|^2$. Consequently, the probability must be continuous at $|c_1|^2$ and $p(|c_1|^2) = |c_1|^2$ even if $|c_1|^2 \notin \mathbb{Q}$.

3. NONCONTEXTUAL PROBABILITY FOR ENTANGLED STATES

Theorem 1 with NCP can be generalized to include also entangled states, bipartite or multipartite systems and quantum measurement devices.

Let us indicate with C the context, including our experimental devices and the environment, and with $|C_i\rangle \in \mathcal{H}^M$ its possible states. Consider again the measurement of the two observables $\hat{A}$ and $\hat{A}'$ introduced in the proof of Lemma 3 in the previous section. The choice of measuring either the eigenvalues $\{a_2, a_3\}$ or $\{a_2', a_3'\}$ together with a_1 corresponds to two different experimental arrangements A and A'.

Suppose that at time t_0 the system is in the initial state $|\psi_{SC}(t_0)\rangle = |\psi_S\rangle|C_0\rangle$. At time t, after the interaction with the measurement device, the entanglement between the system S with the experimental setup C correlates the observable $\hat{A}$ (or $\hat{A}'$) of the system with the pointers of the experimental device. If we measure $\hat{A}$ we have the state:

$$|\psi_{SC}(t)\rangle = c_1|a_1\rangle|C_1\rangle + c_2|a_2\rangle|C_2\rangle + c_3|a_3\rangle|C_3\rangle, \tag{59}$$

while if we measure $\hat{A}'$ we get:

$$|\psi_{SC'}(t)\rangle = c_1|a_1\rangle|C_1'\rangle + c_2'|a_2'\rangle|C_2'\rangle + c_3'|a_3'\rangle|C_3'\rangle. \tag{60}$$

The global wave functions Equations (59) and (60) are different in the different contexts A and A'. So the two probabilities

$$p_A(a_1) = p(|a_1\rangle|C_1\rangle; |\psi_{SC}(t)\rangle),$$
$$p_{A'}(a_1) = p(|a_1\rangle|C_1'\rangle; |\psi_{SC'}(t)\rangle), \tag{61}$$

may be different. The possible contextuality of probability is now included inside the description of the global system $S+C$ of the quantum state. We notice that it is no longer possible to write

$$p_A(a_1) = p(|a_1\rangle|C_1\rangle; |\psi_{SC}(t)\rangle, \{|a_1\rangle, |a_2\rangle, |a_3\rangle\}), \tag{62}$$

if we accept the completeness of QM. Otherwise, indeed, some parameters would be outside the description of the state $|\psi_{SC}(t)\rangle$, which now englobes not only the system but also the context of the measurement. This should imply the existence of hidden variables.

Assuming NCP requires the equality between the probabilities in Equation (61):

$$p(|a_1\rangle|C_1\rangle;|\psi_{SC}(t)\rangle) \equiv p(|a_1\rangle|C_1'\rangle;|\psi_{SC'}(t)\rangle), \tag{63}$$

which generalizes the condition in Equation (23). In fact, Equation (63), by the perfect correlation property of Equation (17), becomes equivalent to Equation (23).

In a similar way as in the previous derivation, by using the state of Equation (59) and a state $|\psi_S\rangle$ with a new basis $\{|a_i'\rangle\}$ such that $c_i' = 0$ for $i \geq 3$, $|a_1'\rangle \equiv |a_1\rangle$ we can write:

$$|\psi_{SC'}(t)\rangle = c_1|a_1\rangle|C_1'\rangle + |c_2'||a_2'\rangle|C_2'\rangle. \tag{64}$$

From the condition of NCP it is not difficult to deduce that $p(|a_i\rangle|C_i\rangle;|\psi_{SC}(t)\rangle) = p(|c_i|)$ must hold. Then, employing states like in Equation (59) and Equation (64) and from NCP, normalization condition and conservation of probability, it is possible to prove again the linearity of the probabilities with respect to the square moduli of the amplitudes (continuity can also be generalized to the present case). Putting all pieces together, we get

$$\begin{aligned} p(|a_i\rangle|C_i\rangle;|\psi_{SC}(t)\rangle) &= |c_i|^2 \\ &= |\langle a_i|\langle C_i||\psi_{SC}(t)\rangle|^2 \\ &= |\langle a_i|\psi_S\rangle|^2. \end{aligned} \tag{65}$$

We note that the state of the global system $S+C$ is in the Hilbert space $\mathcal{H}^N \otimes \mathcal{H}^M$ with dimension $N \cdot M$. If $M \geq 3$ this allows us to include in the derivation of the Born rule also states with $N = 2$. In fact, consider a system $\mathcal{S}$ described by a state $|\psi_S\rangle$ in a Hilbert space with $\dim\mathcal{H}=2$ and basis $\{|a_1\rangle,|a_2\rangle\}$. With a measurement we could use the context C as an ancilla system to build the following entangled states:

$$\begin{aligned}
|\psi_{SC}^{0}\rangle &= c_1|a_1\rangle|C_1\rangle + c_2|a_2\rangle|C_2\rangle, \\
|\psi_{SC'}^{1}\rangle &= c_1|a_1\rangle|C_1'\rangle + c_2'|a_2\rangle|C_2'\rangle + c_3'|a_2\rangle|C_3'\rangle,
\end{aligned} \tag{66}$$

where $|\psi_{SC'}^{1}\rangle$ is obtained by dividing the part of the state $|\psi_S\rangle$ associated with the eigenvalue a_2 into two regions. So it is possible to demonstrate that

$p(a_i)=|\langle a_i|\psi_S\rangle|^2$ as in the previous cases. We note that the condition $\dim\mathcal{H}\geq 3$ for the state of the ancilla system C coupled with S is necessary to obtain, from Equations (66) by the NCP property of Equation (63), a constraint of additivity and linearity similar to that which is at Equations (39)-(46).

4. NONCONTEXTUAL PROBABILITY AND RESOLUTION OF THE MEASUREMENT

A simple application of Lemma 3 shows that equal amplitudes are related to equal probabilities, i.e: if $c_i=c_k$ with $i\neq k$ then $p(a_i)=p(a_k)$. Whereas Lemma 3 is derived from the NCP principle, the same result was obtained by Zurek and Deutsch on the base of two different invariance principles. Then both authors study the more general case with states having different rational amplitudes using the fine-graining technique. We shortly summarize their results. We consider the fine-graining in Zurek's approach with the environment in a simple situation (our remarks also hold for Deutsch's similar derivation). We assume that the state of the system $\mathcal{S}$ is in a Hilbert space with $\dim\mathcal{H}=2$, entangled with the state of the environment $\mathcal{E}$ such that their joint state is

$$|\psi_{\mathcal{S}\mathcal{E}}\rangle = \frac{1}{\sqrt{N}}|a_1\rangle|\varepsilon_1\rangle + \sqrt{\frac{N-1}{N}}|a_2\rangle|\varepsilon_2\rangle. \tag{67}$$

The idea is to transform this state with unequal coefficients into a state with equal coefficients by fine-graining. This goal is reached by extracting from the environment an ancilla system $\mathcal{C}$ having states

$$|C_1\rangle = |C_1'\rangle, \quad |C_2\rangle = \sum_{i=2}^{N}|C_i'\rangle/\sqrt{N-1}, \tag{68}$$

which become correlated with the states $\{|a_1\rangle,|a_2\rangle\}$ of $\mathcal{S}$. Within a quantum measurement, $\mathcal{C}$ can be regarded as a counter and $\{|C_i'\rangle\}$ as a new orthonormal basis of pointer states. Denoting with $\{|e_i\rangle\}$ the new states of the environment without $\mathcal{C}$, after the interaction between $\mathcal{S}$ and $\mathcal{C}$ we have the joint system

$$|\psi_{\mathcal{S}\mathcal{C}\mathcal{E}}\rangle = \frac{1}{\sqrt{N}}|a_1\rangle|C_1\rangle|e_1\rangle + \sqrt{\frac{N-1}{N}}|a_2\rangle|C_2\rangle|e_2\rangle, \tag{69}$$

which can be expressed with the new basis of pointer states of Equation (68):

$$|\psi_{\mathcal{S}\mathcal{C}'\mathcal{E}}\rangle = \frac{1}{\sqrt{N}}|a_1\rangle|C_1\rangle|e_1\rangle + \frac{1}{\sqrt{N}}\sum_{i=2}^{N}|a_2\rangle|C_i'\rangle|e_2\rangle. \tag{70}$$

Now we have a new state with equal amplitudes. But how can we be sure that after fine-graining the probability $p(|a_1\rangle)$ with the state $|\psi_{SC'\varepsilon}\rangle$ is the same with the state $|\psi_{SC\varepsilon}\rangle$ or $|\psi_{S\varepsilon}\rangle$? In principle they could be different, so to legitimately use the fine-graining reasoning we have to assume the equality:

$$p(|a_1\rangle;|\psi_{SC'\varepsilon}\rangle) = p(|a_1\rangle;|\psi_{SC\varepsilon}\rangle), \tag{71}$$

equivalent to the condition of NCP encountered in Equation (63) of Section 3. Then, in their derivations Zurek and Deutsch implicitly assume NCP. We notice that fine-graining is equivalent to a change of measurement resolution, and this has been used in our derivation as well. A change of resolution, for instance, is considered going from Equation (39) to Equation (40). There we assumed from NCP that $p(|a_1\rangle)$ is the same in the two equations and therefore:

$$p(|a_2'\rangle) = p(|a_2\rangle) + p(|a_3\rangle), \tag{72}$$

where $p(|a_2'\rangle)$ is the probability of finding the state of the system in the subspace orthonormal to $|a_1\rangle$, whereas $p(|a_2\rangle)$ and $p(|a_3\rangle)$ are the probabilities of finding it, respectively, in the subspaces $|a_2\rangle$ and $|a_3\rangle$ of the subspace orthonormal to $|a_1\rangle$. In fact, in terms of projectors, Equation (72) can be written as Gleason's Postulate 2:

$$\begin{aligned} &p(|a_2\rangle\langle a_2| + |a_3\rangle\langle a_3|) \\ &= p(|a_2\rangle\langle a_2|) + p(|a_3\rangle\langle a_3|). \end{aligned} \tag{73}$$

Thus NCP could be physically interpreted as a probability invariance when we change the resolution of the measurement connected with other properties of the system. This interpretation gives physical insight to the NCP principle as an invariance under changing resolution. The probability of one measurement outcome should not change if we make coarse or fine-graining over other measurement outcomes. We point out that coarse/fine graining could be done through a post-processing after the first measurement. Therefore, NCP is a natural assumption if we do not want signalling from the future to the past.

5. NONCONTEXTUAL PROBABILITY AND NO-FASTER-THAN-LIGHT SIGNALLING

In general, every measurement in the real world happens in the spacetime. Every measurement of any observable is necessarily reduced to a measurement of position at a certain time. For instance, the Stern-Gerlach apparatus correlates a spin observable with a position observable so that different values of the spin univocally correspond to different position values of the system. A measurement of position detects the spin value of the system.

Let us introduce in our representation of S the observable position $\hat{X}$. We assume that at time t_0, the system is in the initial state $|\psi_{SC}(t_0)\rangle = |\psi_S\rangle|x_0\rangle|C_0\rangle$. For simplicity of notation, from now on we denote with $|C_{xi}\rangle = |x_i\rangle|C_i\rangle$ the position state of the system with its context, where the letter x will remind us the dependence on the position of the system and the experimental apparatus. Suppose that we can measure either the values of the observable $\hat{A}$ or the values of the observable $\hat{A}'$. In the context A of measurement we get:

$$|\psi_{SC}(t)\rangle = c_1|a_1\rangle|C_{x1}\rangle + c_2|a_2\rangle|C_{x2}\rangle + c_3|a_3\rangle|C_{x3}\rangle, \quad (74)$$

while for the experimental context A' we have the different state:

$$|\psi_{SC'}(t)\rangle = c_1|a_1\rangle|C'_{x1}\rangle + c'_2|a'_2\rangle|C'_{x2}\rangle + c'_3|a'_3\rangle|C'_{x3}\rangle. \quad (75)$$

We want to prove the following

Theorem 2:

If we assume the no-faster-than-light signalling condition for measurement events which are spacelike separate, then the probability $p(a_i)$ must be noncontextual.

Proof:

Consider a set of identically prepared systems S in the initial state $|\psi_{SC}(t_0)\rangle$. In order to estimate the probabilities $\{p(a_1), p(a_2), p(a_3)\}$ and $\{p(a_1), p(a'_2), p(a'_3)\}$, let us assume that our experimental device is arranged to measure how many systems have the eigenvalue a_1 at the point x_1 and the eigenvalues $\{a_2, a_3\}$ (Equation (74)) or the eigenvalues $\{a'_2, a'_3\}$ (Equation (75)) at the points x_2 and x_3. This could be done by first dividing the systems S in two parts, one associated with the state $|a_1\rangle$ at the point x_1 and an other one associated with the orthogonal state $|a_{1\perp}\rangle$ at a different point $x_{2\perp}$ (see the superposition state in Equation (78)), then by dividing such a state in two sub-parts, associated either with the basis $\{|a_2\rangle,|a_3\rangle\}$, or with the basis $\{|a_2'\rangle,|a_3'\rangle\}$ at the points x_2, x_3, respectively connected with the states in Equations (74)-(75).

If the measurements of $\{a_2,a_3\}$ or $\{a_2',a_3'\}$ at the points x_2 and x_3 are events spacelike separate with respect to the measurement of a_1 at x_1, according to the relativistic hypothesis of no-signalling, the probability $p(a_1)$ of getting the eigenvalue a_1 has to be the same in both different experimental contexts A and A'. Otherwise an observer situated in the place of measurement of $\{a_2,a_3\}$ or $\{a_2',a_3'\}$ would be able to send a superluminal signal to another observer situated in the place where a_1 is measured. Then, according to the no-faster-than-light signalling hypothesis:

$$p\left(|a_1\rangle|C_{x1}\rangle;|\psi_{SC}(t)\rangle\right)=p\left(|a_1\rangle|C'_{x1}\rangle;|\psi_{SC'}(t)\rangle\right), \tag{76}$$

which is to say, with the result of Equation (17):

$$p(a_1)=p\left(|a_1\rangle;|\psi_{SC}(t)\rangle\right)=p\left(|a_1\rangle;|\psi_{SC'}(t)\rangle\right). \tag{77}$$

Therefore $p(a_1)$ must be noncontextual.

Consider, beside the measurement of the eigenvalues $\{a_1,a_2,a_3\}$ of the observable $\hat{A}$, also the simple possibility of measuring whether our systems assume the eigenvalue a_1 or not, which is equivalent to check whether the systems S are in the state $|a_1\rangle$ or in the subspace $|a_{1\perp}\rangle$ orthogonal to it. In such a case, during the process of measurement, the systems $S+C$ will be in the state:

$$\left|\psi_{SX_\perp}(t)\right\rangle = c_1\left|a_1\right\rangle\left|C_{x1}\right\rangle + c_{2\perp}\left|a_{1\perp}\right\rangle\left|C_{x2\perp}\right\rangle. \tag{78}$$

If the two choices of measuring either $\{|a_1\rangle, |a_2\rangle, |a_3\rangle\}$ or $\{|a_1\rangle, |a_{1\perp}\rangle\}$ correspond to events spacelike separate, as mentioned above, from the condition of no-signalling we must have the NCP condition:

$$p(a_1) = p\left(|a_1\rangle; |\psi_{SC}(t)\rangle\right) = p\left(|a_1\rangle; |\psi_{SC_\perp}(t)\rangle\right). \tag{79}$$

From Equation (79) the conservation of probability imposes also the equation:

$$p\left(|a_{1\perp}\rangle; |\psi_{SC_\perp}(t)\rangle\right)$$
$$= p\left(|a_2\rangle; |\psi_{SC}(t)\rangle\right) + p\left(|a_3\rangle; |\psi_{SC}(t)\rangle\right). \tag{80}$$

As we showed at the end of Section IV, Equation (80) is equivalent to the initial hypothesis adopted by Gleason. Moreover by examining Equation (79) we understand that it is analogous to assuming an invariance condition of the probability under a resolution change of our measurement, or fine/coarse graining. So, at least for spacelike separate measurements, NCP could be deduced by a relativistic principle. However, the Born rule holds for general events, not only for spacelike separate ones. Hence NCP or, equivalently, the invariance of probability when we change resolution, seem to be more general principles.

6. CONCLUDING REMARKS

We have given a new derivation of the Born rule based on the noncontextuality for projective measurements and on some non-statistical postulates of QM. The introduction of a pure quantum state of the system has allowed a more elementary proof than Gleason's theorem. This may seem, at first sight, a loss of generality with respect to Gleason's proof, however Gleason also introduces vectors of basis which can be regarded as states of the system. Moreover, the step from pure states to density operators and to the trace rule for mixed ensembles is quite natural when we already have the Born rule [36].

As in Gleason's, our derivation holds for $\dim\mathcal{H} \geq 3$, where $\dim\mathcal{H}$ could be finite or infinite. The origin of this dimensional limitation lies in the fact that for $\dim\mathcal{H} \geq 3$ the same state vector may belong to several distinct orthonormal bases. This, with NCP, the normalization of the probability and the quantum state, brings us to a constraint of functional linearity between the probability and the amplitude square modulus. Our theorem can be simply generalized in order to include multipartite states. In such a case, if a part of the global system has $\dim\mathcal{H} \geq 3$, the Born rule can also be proved to hold in subspaces with.

We have also given a derivation that the probability is continuous with respect to the amplitudes using measurement noncontextuality. This fills a gap of Zurek's and Deutsch's demonstrations.

Physical motivation for NCP could come from a relativistic principle of no-faster-than-light signalling. However, the Born rule also holds for measurements which are non-spacelike separate, hence NCP seems to be a more fundamental principle. Finally, we have given an interpretation of measurement noncontextuality as an invariance of the probability under a resolution change of our measurement.

ACKNOWLEDGEMENTS

This work has been supported by the EU-STREP Project QIBEC within the activities of the BEC center. QSTAR is the MPQ, IIT, LENS, UniFi joint center for Quantum Science and Technology in Arcetri.

REFERENCES

1. J. A. Wheeler and W. H. Zurek, Eds., "Quantum Theory and Measurement," Princeton University Press, Princeton, 1963.

2. J. von Neumann, "Mathematical Foundations of Quantum Mechanics," Princeton University Press, Princeton, 1955.

3. T. F. Jordan, "Linear Operators for Quantum Mechanics," John Wiley Sons, New York, 1969.

4. A. M. Gleason, "Measures on the Closed Subspaces of a Hilbert Space," Journal of Mathematics and Mechanics, Vol. 6, No. 6, 1957, pp. 885-893.

5. H. Everett, "Relative State Formulation of Quantum Mechanics," Reviews of Modern Physics, Vol. 29, No. 3, 1957, pp. 454-462.

6. B. S. DeWitt, "The Many-Universes Interpretation of Quantum Mechanics," In: B. d'Espagnat, Ed., Foundations of Quantum Mechanics, Academic Press, New York, 1971.

7. D. Finkelstein, "The Logic of Quantum Mechanics," Transactions of the New York Academy of Sciences, Vol. 25, No. 6, 1965, pp. 621-637.

8. J. B. Hartle, "Quantum Mechanics of Individual Systems," American Journal of Physics, Vol. 36, No. 8, 1968, pp. 704-712.

9. N. Graham, "The Measurement of Relative Frequency," In: B. S. DeWitt and N. Graham, Eds., The Many-Worlds Interpretation of Quantum Mechanics, Princeton University Press, Princeton, 1973.

10. R. Geroch, "The Everett Interpretation," Noûs, Vol. 18, No. 4, 1984, pp. 617-633.

11. E. Farhi, J. Goldstone and S. Gutmann, "How Probability Arises in Quantum Mechanics," Annals of Physics, Vol. 192, No. 2, 1989, pp. 368-382.

12. H. Stein, "The Everett Interpretation of Quantum Mechanics: Many Worlds or None?" Noûs, Vol. 18, No. 4, 1984, pp. 635-652.

13. A. Kent, "Against Many-Worlds Interpretations," International Journal of Modern Physics A, Vol. 5, No. 9, 1990, pp. 1745-1762.

14. T. Endo, "Verification of Born's Rule by a Quantum Mechanical Meter," Physics Letters A, Vol. 308, No. 4, 2003, pp. 256-258.

15. C M. Caves, C. A. Fuchs and R. Schack, "Quantum Probabilities as Bayesian Probabilities," Physical Review A, Vol. 65, No. 2, 2002, pp. 1-6.

16. M. Redhead, "Incompleteness, Nonlocality and Realism," Claredon Press, Oxford, 1989.

17. R. W. Spekkens, "Contextuality for Preparations, Transformations, and Unsharp Measurements," Physical Review A, Vol. 71, No. 5, 2005, pp. 1-17.

18. W. H. Zurek, "Environment-Assisted Invariance, Entanglement, and Probabilities in Quantum Physics," Physical Review Letters, Vol. 90, No. 12, 2003, pp. 1-4.

19. M. Schlosshauer and A. Fine, "On Zurek's Derivation of the Born Rule," Foundations of Physics, Vol. 35, No. 197, 2005, pp. 197-213.

20. D. Deutsch, "Quantum Theory of Probability and Decisions," Proceedings of the Royal Society of London A, Vol. 455, No. 1988, 1999, pp. 3129-3137.

21. W. H. Zurek, "Probabilities from Entanglement, Born's Rule from Envariance," Physical Review A, Vol. 71, 2005, pp. 1-29.

22. W. H. Zurek, "Entanglement Symmetry, Amplitudes, and Probabilities: Inverting Born's Rule," Physical Review Letters, Vol. 106, No. 25, 2011, pp 1-4.

23. C. M. Caves, "Notes on Zurek's Derivation of the Quantum Probability Rule," 2005. http://info.phys.unm.edu/caves/reports/ZurekBornderivation.pdf

24. U. Mohrhoff, "Probabilities from Envariance?" International Journal of Quantum Information, Vol. 2, No. 2, 2004 pp. 221-230.

25. H. Barnum, "No-Signalling-Based Version of Zurek's Derivation of Quantum Probabilities: A Note on Environment-Assisted Invariance, Entanglement, and Probabilities in Quantum Physics," 2003. arXiv: quant-ph/0312150

26. H. Barnum, C. M. Caves, J. Finkelstein, C. A. Fuchs and R. Schack, "Quantum Probability from Decision Theory?" Proceedings of the Royal Society of London A, Vol. 456, No. 1997, 2000, pp. 1175-1182.

27. S. Saunders, "Derivation of the Born Rule from Operational Assumptions," Proceedings of the Royal Society of London A, Vol. 460, No. 2046, 2004, pp. 1-18.

28. D. Wallace, "Everettian Rationality: Defending Deutsch's approach to probability in the Everett interpretation," Studies in History and Philosophy of Modern Physics, Vol. 34, No. 3, 2003, pp. 415-438.

29. W. H. Zurek, "Decoherence, Einselection, and the Quantum Origins of the Classical," Reviews of Moder Physics, Vol. 75, No. 3, 2003, pp. 715-775.

30. W. H. Zurek, "Quantum Darwinism and Envariance," 2003. arXiv: quant-ph/0308163

31. C. Cohen-Tannoudji, B. Diu and F. Laloë, "Quantum Mechanics," Hermann and Wiley, Paris, 1977.

32. N. Zettili, "Quantum Mechanics, Concepts and Applications," Wiley, Chichester, 2001.

33. W. Greiner, "Quantum Mechanics, An Introduction," Springer-Verlag, Berlin, 2000.

34. A. Galindo and P. Pascual, "Quantum Mechanics I," Springer-Verlag, Berlin, 1990.

35. M. Kuczma, "An Introduction to the Theory of Functional Equations and Inequalities," Birkhäuser, Basel, 2009.

36. J. J. Sakurai, "Modern Quantum Mechanics," Addison-Wesley, San Francisco, 1994.

CHAPTER 2

Numerical Procedures for Calculating the Probabilities of Recurrent Runs

M. P. Silverman

Department of Physics, Trinity College, Hartford, USA

ABSTRACT

Run count statistics serve a central role in tests of non-randomness of stochastic processes of interest to a wide range of disciplines within the physical sciences, social sciences, business and finance, and other endeavors involving intrinsic uncertainty. To carry out such tests, it is often necessary to calculate two kinds of run count probabilities: 1) the probability that a certain number of trials results in a specified multiple occurrence of an event, or 2) the probability that a specified number of occurrences of an event take place within a fixed number of trials. The use of appropriate generating functions provides a systematic procedure for obtaining the distribution functions of these probabilities. This paper examines relationships among the generating functions applicable to recurrent runs and discusses methods, employing symbolic mathematical software, for implementing numerical extraction of probabilities. In addition, the asymptotic form of the cumulative distribution function is derived, which allows accurate runs statistics to be obtained for sequences of trials so large that computation times for extraction of this information from the generating functions could be impractically long.

KEYWORDS

Recurrent Events; Theory of Runs; Time Series Analysis; Generating Functions; Probability Distributions

1. INTRODUCTION

1.1. Runs Tests for Non-Randomness

A stochastic process generates random outcomes in time or space. Such processes occur widely in the physical and social sciences, as well as in purely practical human activities such as finance, manufacturing, and commerce. Despite their random occurrence—indeed, precisely because of it—the outcomes of a stochastic process will display ordered patterns which a statistically naïve observer may mistakenly interpret as predictively useful information. In recent years, controversial issues over the information content of time series have arisen in a variety of disciplines such as nuclear physics [1] and econophysics (i.e. dynamics of the stock market) [2]. Although it is not possible to prove with certainty that a particular process is random, there are various statistical tests to demonstrate within specified confidence limits that it is not random. Among these, nonparametric runs tests are especially useful, in part because of their ease of implementation and statistical power [3].

1.2. Exclusive Runs

An exclusive run is an unbroken sequence of similar events, ordinarily of a binary nature. For example, a sequence of symbols aabbbaa comprises 2 runs of a's of length 2 and 1 run of b's of length 3. If the events a and b occur with fixed probabilities throughout the sequence, the stochastic process is of the Bernoulli kind, and the distribution theory of binary runs [4] can be used to test for non-randomness in permutational ordering of any such empirical sequence of outcomes.

It is not necessary for a stochastic process to generate binary events in order to be analyzed for runs. For example, a sequence of n different observations $x_1, x_2, \cdots, x_n$ of a continuous random variable will yield $n-1$ sequential differences that are either positive (+) or negative (?) and therefore again subject to binary runs analysis. The binary elements, however, are not Bernoulli variates since the probability of obtaining an element decreases with its position in the unbroken sequence. Nevertheless, one can test for non-randomness in permutational ordering with a different distribution theory [5].

Although developed initially for testing quality control in manufacturing, exclusive runs and up-down runs have been employed in analysis of a variety of experiments to test the fundamental prediction of quantum mechanics that transitions between quantum states occur randomly [6,7]. A problematic issue in the counting of exclusive or up-down runs is that the length of a run can be changed by future events. Thus, in the succession aabbbaa, the second run of 2 a could change to a run of 3 a or 4 a if the next two trials resulted in ab or aa respectively.

1.3. Runs of Recurrent Events

A third kind of runs analysis, based on Feller's theory of recurrent events [8], was recently employed to examine certain quantum optical processes for evidence of non-random behavior [9]. A recurrent run of length t, as defined by Feller, is a sequence of non-overlapping, uninterrupted successions of exactly t elements of the same kind. It is distinguished from the other two kinds of runs in that the concept of run length is so defined as to be independent of subsequent trials. For example, in the sequence aaaabaaaaaa, there are two runs of length 4 $[aaaa\,|\,b\,|\,aaaa\,|\,aa]$, three runs of length 3 $[aaa\,|\,ab\,|\,aaa\,|\,aaa]$, and five runs of length 2 $[aa\,|\,aa\,|\,b\,|\,aa\,|\,aa\,|\,aa]$. (Analyzed in terms of exclusive runs, there would have been 1 run of a of length 4 and 1 run of a of length 6, provided the sequence ended at the 11th trial). In a sequence of Bernoulli trials, a recurrent run of length t occurs at the nth trial if the nth trial adds a new run to the sequence. Thus, the recurrent runs of length 4 occur at positions 4, 9, and the recurrent runs of length 3 occur at positions 3, 8, 11.

The advantage of Feller's definition is that runs of a fixed length become recurrent events, and the statistical theory of recurrent events can then be applied to testing empirical data sequences for permutational invariance over a much wider variety of patterns than just those of unbroken sequences of identical binary elements. For example, one may be interested in testing the recurrence of a pattern abab, which, in a quantum optics experiment, might correspond to a sequence of alternate detections of left and right circularly polarized photons, or, in a series of stock price variations, might correspond to a sequence of alternative observations of rising and falling closing prices. Besides the application to runs, the same theoretical foundation may be applied to recurrent events in other forms such as return-to-origin problems, ladder-point problems (instances where a sum of random variables exceeds all preceding sums), and waiting-time problems.

The theory of recurrent runs, the relevant parts of which are examined in the following section, leads to generating functions from which the probability of a run of defined events of specified length can in principle be calculated exactly. As a practical matter, the extraction of these probabilities requires geometrically longer computation times with increasing sequence length. The availability of fast lap-top computers with large random access memory and of symbolic mathematical software of hitherto unparalleled ability to execute series expansions and perform differentiation and integration provides the analyst with computational power unimaginable to the creators of the statistical theory of runs. I report here mathematical strategies for reducing significantly the computation time for the probability of the widely applicable case of k occurrences of runs of length t in a Bernoulli sequence of length n.

2. THEORY AND IMPLEMENTATION OF RECURRENT RUNS

2.1. Probability Generating Functions

Following Feller, I define the recurrent event E to be a run of successes of length t in a sequence of Bermoulli trials with p the probability of a single successful outcome and $q=1-p$ the probability of failure. Consider the following random variables:

$$T_k = \left[\text{number of trials ("waiting time") between } (k-1)^{th} \text{ and } k^{th} \text{ occurrence of E}\right] + 1$$

(1)

$$S_k = \sum_{j=1}^{k} T_j = \text{number of trials up to and including } k^{th} \text{ occurrence of E}$$

(2)

$$N_n = \text{number of occurrences of E in } n \text{ trials}$$ (3)

The distribution of the variable T is defined by

$$\Pr(T = n) \equiv f_n \quad \text{with } f_0 = 0$$ (4)

where f_n is the probability that E occurs for the first time at the nth trial. The generating function of the probabilities of first occurrence is expressed by

$$F(s) = \sum_{n=0}^{\infty} f_n s^n$$ (5)

The number of trials to the kth occurrence of E is then characterized by the random variable S_k in (2), which is a sum of the waiting times of k independent trials, from which it is follows that the associated generating function takes the form

$$F^{(k)}(s) = \sum_{n=0}^{\infty} f_n^{(k)} s^n = \left[F(s)\right]^k = \left(\sum_{n=0}^{\infty} f_n s^n\right)^k$$ (6)

where

$$\Pr\left(S_k = n\right) \equiv f_n^{(k)} \tag{7}$$

is the probability that the kth occurrence of E first takes place at the nth trial.

I leave to the cited literature the proof that the generating function (5) for runs of length t with individual probability of success p is given by

$$F\left(s,p,t\right) \equiv F\left(s\right) = \frac{p^t s^t \left(1-ps\right)}{1-s+qp^t s^{t+1}} \tag{8}$$

from which the mean and standard deviation of the recurrence times follow by differentiation

$$\mu\left(p,t\right) = \left.\frac{\mathrm{d}F\left(s\right)}{\mathrm{d}s}\right|_{s=1} = \frac{1-p^t}{qp^t} \quad \left(q=1-p\right) \tag{9}$$

$$\sigma\left(p,t\right) = \sqrt{\left[\frac{\mathrm{d}^2F\left(s\right)}{\mathrm{d}s^2} - \left(\frac{\mathrm{d}F\left(s\right)}{\mathrm{d}s}\right)^2 + \frac{\mathrm{d}F\left(s\right)}{\mathrm{d}s}\right]_{s=1}} = \left(\left(qp^t\right)^{-2} - \left(2t+1\right)\left(qp^t\right)^{-1} - pq^{-2}\right)^{1/2} \ . \tag{10}$$

For economy of expression, the parameters p, t will be suppressed in the arguments of $F\left(s,p,t\right)$, $\mu\left(p,t\right)$, and $\sigma\left(p,t\right)$ unless needed to avoid ambiguity. In general, these parameters will be chosen and fixed at the outset of any illustrative applications. Note, too, that to obtain a statistical moment from a probability generating function (pgf), the derivatives are evaluated at $s=1$, which leads to a sum of terms, whereas to obtain a probability the derivatives are evaluated at $s=0$, which leads to a single term.

For many applications the analyst's interest is not necessarily in the recurrence time (i.e. number of trials) to the kth occurrence of E, but in the probability that E occurs k times in a fixed number n of trials. The relation connecting the two variates is

$$\Pr(N_n \geq k) = \Pr(S_k \leq n) \qquad (11)$$

The probability $p_{n,k}$ that k events E occur in n trials is then expressible as

$$p_{n,k} = \Pr(N_n = k) = \Pr(S_k \leq n) - \Pr(S_{k+1} \leq n) \qquad (12)$$

and serves in the construction of two pgf's

$$G(z,n) = \sum_{k=0}^{\infty} p_{n,k} z^k \qquad (13)$$

and

$$Q(s,k) = \sum_{n=1}^{\infty} p_{n,k} s^n = \frac{F(s)^k [1 - F(s)]}{1-s} \qquad (14)$$

Note that the summation in (13) is over the number of occurrences, whereas the summation in (14) is over the number of trials. The second equality in (14) follows directly from Equation (11). Multiplying both sides of (13) by s^n and summing over n leads to the bivariate generating function

$$H(s,z) \equiv \sum_{n=1}^{\infty}\left[\sum_{k=0}^{\infty} p_{n,k} z^k\right] s^n = \frac{1-F(s)}{(1-s)(1-zF(s))} \tag{15}$$

from which the probabilities $p_{n,k}$ are calculable by series expansion of both sides of the equality.

A sense of the structure of the formalism can be obtained by considering the case of recurrent runs of length $t=3$ for a stochastic process with $p=\frac{1}{2}$. Substitution of these conditions into Equation (8) for $F(s)$ yields the following rational expression for the right side of Equation (15) and its corresponding Taylor-series expansion

$$G(s,z) = \frac{-2\left(s^2+2s+4\right)}{\left(s^3+zs^3+2s^2+4s-8\right)}$$
$$= 1+s+s^2+\left(\frac{1}{8}z+\frac{7}{8}\right)s^3+\left(\frac{3}{16}z+\frac{13}{16}\right)s^4+\left(\frac{1}{4}z+\frac{3}{4}\right)s^5+O\left(s^6\right) \tag{16}$$

to order s^6. Recall that the powers of s designate the number of trials, and the powers of z designate the number of recurrences of runs of length 3. For a fixed power of s, the sum of the coefficients of the powers of z within each bracketed expression equal unity, as they must by the completeness relation for the probability of mutually exclusive outcomes. Note that the first three terms $\left(s^0+s^1+s^2\right)$ are independent of z—i.e. contain only powers z^0—since there cannot be runs of length 3 in a sequence of no more than 2 trials. For 3 trials, the probability of 0 runs of length 3 is 7/8 and the probability of 1 run of length 3 is 1/8. For 5 trials, however, the probability of 0 runs is 1/4 and the probability of 1 run is 3/4. This pattern persists: (a) to obtain a run of length t, the sequence of trials must be of length $s \geq t$, and (b) the greater the number of trials, the higher is the probability of obtaining longer runs.

2.2. Moment Generating Functions

It is not necessary to know the individual $p_{n,k}$ to determine the mean number of recurrent runs

$$\langle N_n \rangle = \sum_{k-0}^{\infty} k p_{n,k} \tag{17}$$

which, for many applications in the physical sciences and elsewhere, is the experimentally observed quantity of interest. Multiplying both sides of Equation (17) by s^n and summing n over the range $(1,\infty)$ leads to the generating function for the distribution of $\langle N_n \rangle$

$$M_1(s) = \sum_{n=1}^{\infty} \langle N_n \rangle s^n = \frac{F(s)}{(1-s)(1-F(s))}. \tag{18}$$

Starting from the relation

$$\langle N_n^2 \rangle = \sum_{k-0}^{\infty} k^2 p_{n,k} \tag{19}$$

and following the same procedure that led to (18) yields the generating function for the distribution $\langle N_n^2 \rangle$

$$M_2(s) = \sum_{n=1}^{\infty} \langle N_n^2 \rangle s^n = \frac{F(s) + F(s)^2}{(1-s)(1-F(s))^2} . \tag{20}$$

From the generating functions (18) and (20) one can deduce the asymptotic relations for mean and variance

$$\langle N_n \rangle \approx n\mu^{-1} . \tag{21}$$

and

$$\operatorname{var}(N_n) \approx n\sigma^2 \mu^{-3} . \tag{22}$$

2.3. Numerical Procedures

The statistics (probabilities and expectation values) for any physically meaningful choice of probability of success p, run length t, and number of trials n are deducible exactly from expressions (15) and (18) in the manner previously illustrated. For many applications, however, particularly where it is possible to accumulate long sequences of data as is often the case in atomic, nuclear and elementary particle physics experiments or investigations of stock market time series, the tests for evidence of non-random behavior are best made by examining long runs. Suppose, for example, one wanted the probability of obtaining the number of occurrences of runs of length 50 in a sequence of 100 trials. One approach, leading directly to all non-vanishing probabilities, would be to extract the 100th term

$$p_{100,50} = \frac{1}{562949953421312}\left[\left(\frac{1125899906842623}{2251799813685248} + \frac{1}{2251799813685248}z\right)(z+1)\right]$$
$$\approx 1.0000 + 2.3093\times10^{-14}z + 7.8886\times10^{-31}z^2$$

from the series expansion of the bivariate generating function (15). Powerful symbolic mathematical software such as Maple or Mathematica permits one to do this up to a certain order limited by the speed and memory of one's computer, but these calculational tools may become insufficient when one is seeking exact probabilities of runs in data sequences of thousands to millions of bits.

Using complex variable theory, one can extract expressions for $p_{n,k}$ and $\langle N_n \rangle$ by evaluation of contour integrals

$$\begin{aligned} p_{n,k} &= \frac{1}{2\pi i}\oint_C P_n(\xi)\xi^{-(k+1)}d\xi \\ &= \left(\frac{1}{2\pi i}\right)^2 \oint_C d\zeta \frac{1-F(\zeta)}{(1-\zeta)\zeta^{n+1}} \oint_C \frac{d\xi}{(1-\xi F(\zeta))\xi^{k+1}} \\ &= \frac{1}{n!}\left\{\frac{d^n}{ds^n}\left[(1-s)^{-1}(1-F(s))F(s)^k\right]\right\}\Big|_{s=0} \end{aligned} \tag{23}$$

and similarly

$$\langle N_n \rangle = \frac{1}{2\pi i}\oint_C M_1(\xi)\xi^{-(k+1)}d\xi = \frac{1}{n!}\left\{\frac{d^n}{ds^n}\left[(1-s)^{-1}(1-F(s))^{-1}F(s)\right]\right\}\Big|_{s=0} \tag{24}$$

where C is the unit circle and the generating function $F(s)$, given by Equation (8), specifies the single-event probability p and run length t. Contrary to first impression, however, the execution of expressions

(23) or (24) for $p_{n,k}$ or $\langle N_n \rangle$ directly by differentiation, instead of by a series expansion of the corresponding generators up to the order that yields the desired $p_{n,k}$ or $\langle N_n \rangle$, is not computationally economic. The computer, in fact, executes the series expansion of the generator much more rapidly than it performs symbolic differentiation.

Because the generator $M_1(s)$ [Equation (18)] is univariate, one can take advantage of the rapidity with which symbolic mathematical software executes a series expansion to obtain exact values $\langle N_n \rangle$ for very long sequence lengths by the following simple procedure:

1) For given p and t, express $M_1(s)$ as a rational function $g(s)/h(s)$ of s.

2) Convert $g(s)/h(s)$ to a power series $\sum_{j=0}^{\infty} \langle N_j \rangle s^j$. In Maple this can be done by the command

$convert(\cdots, FormalPowerSeries, s)$ where the ellipsis $(\cdots)$ represents either the expression to be converted or the Maple equation number of that expression.

3) Extract the single desired term $\langle N_n \rangle$. In Maple this can be done by filtering the sum: $\langle N_n \rangle = \sum_{j=n}^{n} \langle N_j \rangle s^j$.

Alternatively, one can convert the series generated to order $O(n+1)$ to a polynomial of degree n, and use the command lcoeff to extract the leading coefficient of the polynomial.

As an illustration, consider the calculation (by means of Maple) of the exact mean number of runs of length $t=4$ in a sequence of 1 million trials with probability of success $p=0.5$. Following the foregoing steps, we have

1) $$M_1(s)=\frac{F(s)}{(1-s)(1-F(s))}=\frac{1}{2}\frac{s^4}{(s^4+s^3+2s^2+4s-8)(s-1)}$$

2) $$convert(M_1(s),FormalPowerSeries,s)\rightarrow\sum_{j=0}^{\infty}(\cdots)_j s^j$$

3) $$evalf\left(expand\left(\sum_{j=n}^{n}(\cdots)_j s^j\right)\right)\rightarrow N_n s^n$$

where $(\cdots)_j$ is to be replaced by the actual j^{th} numerical element in the sum obtained in step (2). The arrow, inserted by the author, symbolically points to the form of the output. Extraction of this element for the specified conditions led to $\langle N_{1,000,000}\rangle=33,333.258$ in a fraction of a second. In Maple, the command evalf calls for numerical evaluation of expressions; omitting this command results in an exact fraction, which for a sequence length of 1 million is too unwieldy to be useful. Use of the command lcoeff yields the same result, but executes more slowly for large n.

The procedure described above for converting the rational expression of s into a formal power series in s did not work with the bivariate generator $H(s,z)$, which involved a product of z with a power of s (depending on run length t) in the denominator, and required, for the conversion, solution of the roots of a high-order (>2) algebraic equation. Maple did return, however, the recursion relation for the coefficients of the formal power series. An alternative procedure to isolate the values $p_{n,k}$ for fixed n, which still relied on the computational speed of series expansion and worked well for sequence lengths in the thousands, is the following:

1) For given p and t, express $H(s,z)$ as a rational function of s and z.

2) Generate a series expansion of $H(s,z)$ to order n and $n+1$.

3) Convert the series expansions into polynomials $P_{(n+1)}$ and $P_{(n)}$.

4) Subtract one polynomial from the other to obtain an expression of the form

$$P_{(n+1)} - P_{(n)} \rightarrow \left[p_{n,0} + p_{n,1}z + p_{n,2}z^2 + \cdots + p_{n,\left\lfloor \frac{n}{t} \right\rfloor} z^{\left\lfloor \frac{n}{t} \right\rfloor} \right] s^n$$

where $\left\lfloor \frac{n}{t} \right\rfloor$ is the largest integer k such that $kt \le n$. The coefficients $p_{n,k}$ are given as exact fractions.

5) Evaluate the set $\{p_{n,k}\}$ as floating-point numbers, if desired.

As an example, the procedure led in under 10 seconds to the full set $p_{1000,k}\ \{k = 0, \cdots, 200\}$ for the probability of k occurrences of runs of length 4 in a sequence of 1000 trials. The calculations described in this section were performed with a laptop computer (Intel-based Mac Powerbook) running Maple 16.

One final procedure, particularly suitable when only selected probabilities of the full set $\{p_{n,k}\}$ are desired, is to obtain these probabilities directly from the generating function $Q(s,k)$ in (14). As in the previous examples, this can be accomplished in either of two ways: (1) by evaluating the leading coefficient of the polynomial in the desired degree n, or (2) by filtering the power series for the nth term. As an example, consider the probability $P_{100,4}$ for obtaining 4 runs of length $t=4$ in a series of 100 trials with probability of success $p=0.5$. The generating function then takes the form

$$Q(s,k,p,t)=Q(s,4,0.5,4)=-\frac{2s^{16}\left(s^3+2s^2+4s+8\right)}{\left(s^4+2s^3+4s^2+8s-16\right)^5}$$

In Maple, method (1) proceeds as follows:

1) $convert\left(-\frac{2s^{16}\left(s^3+2s^2+4s+8\right)}{\left(s^4+2s^3+4s^2+8s-16\right)^5}, FormalPowerSeries, s\right)$

2) $evalf\left(expand\left(\sum_{k=100}^{100}(\cdots)\right)\right)\rightarrow 0.2007906348s^{100}$

where the ellipsis is to be replaced by the kth term produced in step (1). In using Maple to execute method (2), one proceeds in a single step once the rational expression $Q(s,4,0.5,4)$ has been obtained:

$$evalf\left(lcoeff\left(convert\left(series\left(Q(s,4,0.5,4),s=0,101\right),polynom\right)\right)\right)\rightarrow 0.2007906348$$

Note that the series must be expanded to $n+1$ terms in order to obtain the coefficient of s^n, since the summation index begins at 0.

3. GENERATING FUNCTION OF CUMULATIVE PROBABILITY

For many applications in the physical sciences and elsewhere, the full set of probabilities $\{p_{n,k}\}$ provides more information than is desirable or usable. Moreover, because $p_{n,k}$ for large k may be very small and the variance relatively large, the more observationally stable statistic is the probability of obtaining k or more occurrences of the specified event, or in other words, the complementary cumulative probability (ccp)

$$\Pr(N_n \geq k) \equiv \tilde{P}_{nk} = \sum_{j=k}^{\infty} p_{nj} \tag{25}$$

introduced in Equation (11). Experimental situations calling for preferential usage of a cumulative probability distribution over a probability function abound in the physical sciences, as, for example, in the analysis of fragmentation [10] and other stochastic processes leading to a power-law distribution.

The generating function for the ccp is derivable from $Q(s,k)$ by summing over the recurrence index k

$$C(s,k) \equiv \sum_{n=1}^{\infty} \tilde{P}_{nk} s^n = \sum_{n=1}^{\infty} \left[\sum_{j=k}^{\infty} Q(s,j) \right] s^n = \frac{F(s)^k}{1-s} . \tag{26}$$

(It is understood that $F(s)$, and therefore $C(s,k)$, are also functions of p and t). Thus, one can calculate $\Pr(N_n \geq k)$ directly from the generator (26) by use of the methods previously described. A modern laptop running Maple can return results for

$\Pr(N_n \geq k)$ within minutes for sequence lengths n on the order of many 1000's.

4. ASYMPTOTIC DISTRIBUTIONS

One can show by application of the Central Limit Theorem (CLT) to relation (11) that for sufficiently large number of trials n and number of occurrences k, the number N_n of runs of length t produced in n trials is approximately normally distributed with mean and variance given by relations (21) and (22). The approximation, whose relative accuracy improves in the limit of increasing n, is actually quite good even for modest values of n, as shown in Table 1 for $n = 100$. Expansion of the generating function $M_1(s)$ yielded the exact mean value as an integer or fraction, which was then expressed as a floating-point number to three significant figures for comparison with the Gaussian approximation. It is to be noted from the Table that the Gaussian approximation overstates the mean values, and that the absolute error $\left[p_{n,k}^{\text{Gauss}} - p_{n,k}^{\text{Exact}}\right]$, in contrast to the relative error $\left[\left(p_{n,k}^{\text{Gauss}} - p_{n,k}^{\text{Exact}}\right) / p_{n,k}^{\text{Exact}}\right]$ increases with run length and number of trials.

The Gaussian approximation, however, does not work well in estimating the cumulative probability $\tilde{P}_{nk}$ for large n and long runs. However, one can obtain substantial improvement by taking account of the asymptotic behavior of $\mu(p,t)$ and $\sigma(p,t)$, and the equivalence in Equation (11). Expansion in probability p for fixed run length t of expressions (9) and (10) yields the series

$$\left.\begin{matrix}\mu(p,t)\\ \sigma(p,t)\end{matrix}\right\} \to p^{-t} + p^{-t+1} + \cdots + p^{-1} + \begin{cases} O(p^0) \\ -c(t) + O(p) \end{cases} \quad (27)$$

in which the constant $c(t)$ is orders of magnitude smaller than the leading terms. Thus, in practical terms, $\mu(p,t)$ and $\sigma(p,t)$ are virtually equal for long runs, as illustrated in Table 2 for $p=0.5$.

Recall that $\mu(p,t)$ and $\sigma(p,t)$, which were derived from the generating function $F(s,p,t)$, are respectively the mean number of trials to the first occurrence of event E, which is a run of length t. The equivalence of the mean and standard deviation suggests that the asymptotic distribution of the random variable T in Equation (1) is exponential [11] $E(\lambda)$ with parameter $\lambda = \mu^{-1}$. Then the random variable S_k in Equation (2),

which is a sum of k independent exponential random variables $\sum_{j=1}^{k} T_j$, follows a gamma distribution

$Gam(\lambda,k)$ with cumulative probability function

$$\Pr(S_k \le n) \approx \frac{1}{\Gamma(k)} \int_0^{n\mu^{-1}} x^{k-1} e^{-x} dx \quad (28)$$

where $\Gamma(k)$ is the standard gamma function equal to $(k-1)!$ for integer k.

Table 1. Comparison of exact and Gaussian mean numbers of runs for n = 100 trials with p = 0.5.

Run Length t	Mean Number of Runs (Exact)	Mean Number of Runs (Gaussian Approximation)
1	50	50
2	16.556	16.667
3	7.041	7.143
4	3.258	3.333
5	1.562	1.613
6	7.611 (−1)	7.937 (−1)
7	3.738 (−1)	3.937 (−1)
8	1.842 (−1)	1.961 (−1)
9	9.098 (−2)	9.785 (−2)
10	4.496 (−2)	4.888 (−2)
11	2.222 (−2)	2.443 (−2)
12	1.099 (−2)	1.221 (−2)
13	5.433 (−3)	6.104 (−3)
14	2.686 (−3)	3.052 (−3)
15	1.328 (−3)	1.526 (−3)
16	6.561 (−4)	7.630 (−4)
17	3.243 (−4)	3.815 (−4)
18	1.602 (−4)	1.907 (−4)
19	7.916 (−5)	9.537 (−5)
20	3.910 (−5)	4.768 (−5)
40	2.819 (−11)	4.547 (−11)
60	1.820 (−17)	4.337 (−17)

Table 2. Asymptotic mean and SD of S_1.

Run Length t	$\mu(0.5,t)$	$\sigma(0.5,t)$
1	2.00	1.41
5	62.00	58.22
10	2046.00	2037.47
20	2.09715×10^6	2.09713×10^6
30	2.14748×10^9	2.14748×10^9

By virtue of equivalence (11), Equation (28) also yields a closed-form asymptotic relation for the sought-for cumulative probability distribution $\Pr(N_n \geq k)$.

Consider, for example, the probability of 1 or more runs of length 20 in 2000 trials with individual probability of success 0.5. A comparison of the results of (a) Equation (28), (b) the cumulative Gaussian distribution with mean and variance given by relations (21) and (22), and (c) the exact calculation obtained from the generating function (26)

$$\Pr(N_{2000} \geq 1)\big|_{\text{Gamma Dist}} = 9.532\times10^{-4}$$
$$\Pr(N_{2000} \geq 1)\big|_{\text{Gaussan Dist}} = 1.588\times10^{-1}$$
$$\Pr(N_{2000} \geq 1)\big|_{\text{Exact Generator}} = 9.451\times10^{-4}$$

supports the distribution (28). If the number of trials is increased to 10^6, the Gamma and Gaussian asymptotic distributions lead to comparable results

$$\Pr\left(N_{10^6} \geq 1\right)\Big|_{\text{Gamma Dist}} = 0.3793$$

$$\Pr\left(N_{10^6} \geq 1\right)\Big|_{\text{Gauss Dist}} = 0.3004$$

$$\Pr\left(N_{10^6} \geq 1\right)\Big|_{\text{Exact Generator}} = 0.3793$$

but the former is still superior to the latter when compared to the probability calculated from the exact generating function.

The relation (28), by which one can calculate the cumulative probability $\Pr(N_n \geq k)$ for large values of n, also allows one to calculate the individual probabilities p_{nk} as a function of number of occurrences k through the identity

$$p_{nk} = P(N_n \geq k) - P(N_n \geq k+1) \approx \frac{1}{\Gamma(k)} \int_0^{n\mu^{-1}} x^{k-1} \left(1 - \frac{x}{k}\right) e^{-x} \mathrm{d}x \tag{29}$$

5. CONCLUSIONS

The theory of recurrent runs provides a statistical basis for rejecting the hypothesis that a series of observations (in time or space) are random. This is a matter that often arises in experimental investigations in atomic, optical, nuclear, and elementary particle physics, as well as in other sciences, finance, and commerce, which may entail a very large number—perhaps in the thousands to millions—of trials or observations.

In this paper theoretical and numerical methods based on different generating functions were derived and investigated to determine (a) the probability for k recurrence runs of length t in n Bernoulli trials,

(b) the complementary cumulative probability $\tilde{P}_{nk} = \Pr(N_n \geq k)$, and (c) the mean number of runs $\langle N_n \rangle$.

The methods reported here can be implemented on modern laptop computers running commercially available symbolic mathematical software, such as Maple (which was the application used by the author). Computation times for application of these methods to data sequences up to millions of trials could range from seconds to minutes.

To compute runs statistics for sequences of intermediate to very long trial numbers, the asymptotic distribution for the number of trials up to and including the kth occurrence $(\lfloor n_t' t \rfloor \geq k \geq 1)$ of a specified run length t was derived and found to be a Gamma distribution $Gam(\mu(p,t)^{-1}, k)$ to excellent approximation. In the limit of very large (technically, infinite) n and k, the Central Limit Theorem (CLT) predicts an asymptotic distribution $\Pr(N_n \geq k)$ of Gaussian form. Both the Gamma and Gaussian asymptotic distributions give comparable results under these circumstances, but the Gaussian approximation is less accurate and fails entirely for values of n and k for which the CLT does not apply.

REFERENCES

1. M. P. Silverman and W. Strange, "Search for Correlated Fluctuations in the β+ Decay of Na-22," Europhysics Letters, Vol. 87, No. 3, 2009, pp. 32001-32006.

2. D. Sornette, "Critical Market Crashes," Physics Reports, Vol. 378, No. 1, 2003, pp. 1-98.

3. J. Wolfowitz, "On the Theory of Runs with Some Applications to Quality Control," The Annals of Mathematical Statistics, Vol. 14, No. 3, 1943, pp. 380-288.

4. M. Mood, "The Distribution Theory of Runs," The Annals of Mathematical Statistics, Vol. 11, No. 4, 1940, pp. 367-392.

5. H. Levene and J. Wolfowitz, "The Covariance Matrix of Runs Up and Down," The Annals of Mathematical Statistics, Vol. 15, No. 1, 1944, pp. 58-69.

6. M. P. Silverman, W. Strange, C. Silverman and T. C. Lipscombe, "Tests for Randomness of Spontaneous Quantum Decay," Physical Review A, Vol. 61, No. 4, 2000, Article ID: 042106.

7. M. P. Silverman and W. Strange, "Experimental Tests for Randomness of Quantum Decay Examined as a Markov Process," Physics Letters A, Vol. 272, No. 1-2, 2000, pp. 1-9.

8. W. Feller, "Fluctuation Theory of Recurrent Events," Transactions of the American Mathematical Society, Vol. 67, 1949, pp. 98-119.

9. D. Branning, A. Katcher, W. Strange and M. P. Silverman, "Search for Patterns in Sequences of Single-Photon Polarization Measurements," Journal of the Optical Society of America B, Vol. 28, No. 6, 2011, pp. 1423-1430.

10. M. P. Silverman, W. Strange, J. Bower and L. Ikejimba, "Fragmentation of Explosively Metastable Glass," Physica Scripta, Vol. 85, 2012, Article ID: 065403.

11. M. Mood, F. A. Graybill and D. C. Boes, "Introduction to the Theory of Statistics," 3rd Edition, McGraw-Hill, New York, 1974, pp. 540-541.

CHAPTER 3

Self-Organized Complexity and Coherent Infomax from the Viewpoint of Jaynes's Probability Theory

William A. Phillips [1,2]

[1] Department of Psychology, University of Stirling, Stirling FK9 4LA, UK; Email: Tel.: 44-0-1786-467640; Fax: 44-0-1786-467641

[2] Frankfurt Institute of Advanced Studies, Frankfurt, 60438, Germany

ABSTRACT

This paper discusses concepts of self-organized complexity and the theory of Coherent Infomax in the light of Jaynes's probability theory. Coherent Infomax, shows, in principle, how adaptively self-organized complexity can be preserved and improved by using probabilistic inference that is context-sensitive. It argues that neural systems do this by combining local reliability with flexible, holistic, context-sensitivity. Jaynes argued that the logic of probabilistic inference shows it to be based upon Bayesian and Maximum Entropy methods or special cases of them. He presented his probability theory as the logic of science; here it is considered as the logic of life. It is concluded that the theory of Coherent Infomax specifies a general objective for probabilistic inference, and that contextual interactions in neural systems perform functions required of the scientist within Jaynes's theory.

KEYWORDS

self-organization; complexity; Coherent Infomax; Jaynes; probability theory; probabilistic inference; neural computation; information; context-sensitivity; coordination

1. INTRODUCTION

Many forms of organized complexity have arisen in nature's long journey from uniformity to maximal entropy. On earth, biological systems have created diverse forms of adaptively self-organized complexity despite the ever present forces of noise and disorder. This self-organization occurs in open, holistic, far-from-equilibrium, "non-linear" systems with feedback, which makes them highly diverse and hard to predict. They depend on information about their world and themselves, and this information is used for inference-inferences about distal things from proximal signals, and inferences about the likely consequences of possible activities. Though usually implicit, probabilistic inference is central to such systems because adaptation depends upon information about the conditions to which the systems are adapted. Useful inference is possible because the laws of physics are sufficiently reliable, but the endless variety of individual circumstances and the prevalence of deterministic chaos make many things unpredictable. So, to thrive, biological systems must combine local reliability with holistic flexibility.

These arguments suggest several issues on which we need to make progress. What is self-organized complexity? What are the capabilities and constraints of the various forms of inductive inference, e.g., classical versus Bayesian [1], conscious versus unconscious [2]? How is local reliability combined with holistic flexibility? What is context, and how can it be used? How can the information theory measures that have been applied to these issues be tested, and what do they contribute to our understanding?

Better formalisation of these issues is clearly needed, so I will first give a brief outline of some conceptions of self-organized complexity, and then of the theory of Coherent Infomax which uses information theory measures to formalize these issues [3,4,5,6]. This theory was initially developed as a theory of mammalian neocortex, but here the possibility of a broader relevance is considered. A summary of Jaynes's theory of probability and inference will then be given, followed by discussion relating it to self-organized complexity and then to Coherent Infomax. Finally, objectives of probabilistic inference in self-organized systems will be briefly discussed.

2. ORGANIZED COMPLEXITY

An obvious tension underlies the notion of organized complexity. Complexity can most simply be thought of as the amount of information in a system. That kind of complexity is increased by increasing the number of elements in a system and by increasing their independence. In contrast to that, organisation implies decreasing theirindependence. Shannon's information entropy quantifies the former as it increases with independence. What is referred to here as organized complexity is often referred to simply as "complexity", but I prefer to emphasize the tension by using terminology that makes it explicit. Many ways of quantifying the broad notion of organized complexity have been suggested. They typically combine order (organization/coherence) with disorder (entropy/information), and they often do so using measures of mutual information [7]. These measures are designed to ascribe high complexity to systems of many elements that interact in such a way as to achieve effective integration but without imposing such uniformity that their joint entropy is low.

Self-organisation is emphasized here for two basic reasons. First, it relates to a major dilemma underlying Jaynes's account of inference, i.e., does it or does it not imply someone, such as a scientist, who draws the inferences? Second, Jaynes's views will be related to a theory of the inference that is implicit in biological systems, which I assume to be self-organized.

Self-organisation is also common in inanimate physical systems. Bénard convection is a well-known example, and has been used to extract general principles of self-organisation that apply even to highly evolved biological systems such as the mammalian neocortex [8].

I assume that life is adaptively self-organized complexity. This adaptation, which is achieved by both genetic selection and ontogenetic plasticity, implies the selection and improvement of constructive processes that require a high capacity for information transmission. The window of possibility for life in all conceivable universes seems to be extremely small. Furthermore, if it depends on liquid water, as seems likely, it may also be small in our own actual universe. In addition to being dependent on information, life creates an information explosion [9] because living things are highly diverse, even down to the molecular level, while also being well coordinated both within and across individual organisms. Macroscopic thermodynamic properties can be accurately estimated by averaging over vast numbers of elements that are assumed to be identical and with independent dynamics. Within living things even a single molecule can have significant effects on its macroscopic properties, and the activities of their elements are not independent, but highly interdependent. Their diversity and ability to surprise us is what makes them so interesting.

3. COHERENT INFOMAX

The centrality of probabilistic inference to life is most obvious in neural systems. Helmholtz correctly emphasized the role of unconscious inference in perception, and many demonstrations of this can be given [10]. Friston [11] has now shown formally how such unconscious inference may also be central to both reinforcement learning and motor control, which extends its relevance to much of neural function. Neural systems will therefore be of particular relevance to the following discussions.

The apparent conflict between the requirements of local reliability and holistic flexibility has been prominent in the history of neuroscience, with one or the other being dominant at different times [12]. The perspective outlined here shows how these two requirements are not necessarily mutually exclusive, as has long been assumed, but can be mutually supportive. Its central hypothesis is that there are two classes of neuronal interaction: driving interactions that specify the information content of the output signals to be transmitted by the local neuronal processor, and coordinating contextual interactions that modulate the gain and timing of response to the driving inputs. This theory shows how contextual inputs can disambiguate local signals and dynamically group them into coherent subsets. It is based upon much detailed psychophysical, neurobiological, and clinical evidence [13,14,15].

The great volume and diversity of that evidence suggests that formal clarification of the role of context-sensitivity in probabilistic inference would be a major advance.

Our contribution to this effort has produced the theory of Coherent Infomax [3,4,5,6]. For full formal presentations see the original publications; only a brief outline is given here. It is closely related to several other influential theories [11,16,17,18,19,20,21,22,23]. Though unification of all these theories is a task for the future, it seems feasible. The theory of Coherent Infomax is founded on an objective function that specifies the goal toward which both the short-term dynamics and the long-term plasticity of local neuronal processing elements are directed. Minimally, the function of local processors is to select and compress that information in their primary driving input that is relevant to the current task and situation, as indicated by the contextual input. This is formalized in information theoretic terms as an objective function describing the signal processing work to be done. In short, the goal is to maximize the information transmitted about the driving inputs, giving priority to the three-way mutual information between output, driving, and contextual inputs while minimizing the information transmitted specifically about the context. This objective therefore uses context to control the gain of response to the driving inputs. To show how that objective could be met in neural systems, a biologically plausible activation function for idealized local neural processors was formulated to include the required gain-control, and a learning rule for modifying the synaptic strengths of the connections between these local processors was derived analytically from the objective function [3,4,5]. The learning rule derived has much in common with synaptic plasticity that has been independently discovered using neurophysiological techniques [13].

Endlessly many system architectures can be constructed from such local processing elements. A system architecture suggested by the anatomy of mammalian neocortex is that of at most a few tens of hierarchical layers of processing, with many specialized but interactive local processors at each stage [13]. Feedforward connections between layers are driving, whereas lateral and feedback connections provide coordinating gain-control. Though it is the dynamics of local processors that are specified by Coherent Infomax, its objective requires them to coordinate their activities with other local processors, thus producing patterns of activity that tend to maximize overall holistic coherence. This shows, in principle, how neural systems can use the information that is available to them to perform probabilistic inference in a way that combines local reliability with holistic flexibility.

The Coherent Infomax objective can be seen as a form of statistical latent structure analysis [3,4,5,6]. Its goal is to discover variables defined upon each of many distinct datasets that have predictive relations with variables defined upon other datasets. In short, its goal is to discover what predicts what. Therefore it concurrently discovers related variables and their relations; the variables do not need to be known in advance. So far, it has been mostly studied in relation to perceptual data-processing tasks, which has the advantage of drawing attention to the fact that inference is not at all restricted to the prediction of future events. It can equally well apply across space, or backwards in time. In visual perception, predictive relationships between concurrent datasets can be used to resolve local ambiguities within them. In auditory perception, inferences from later inputs can similarly be used to resolve ambiguities in earlier inputs. Current astronomical data can be used to predict events in the distant past.

The objective of Coherent Infomax is related to conceptions of organized complexity, such as "effective complexity", though it was not derived from them. The contextual interactions central to the theory maximize organized complexity because they coordinate activities while not becoming confounded with the information that those activities variously transmit. Furthermore, Coherent Infomax is highly compatible with the small-world network architectures conducive to high complexity on these measures [7]. It assumes a system composed of many local processors with logically independent inputs, and with dynamics that allows them to communicate while preserving their distinct identities. This communication discovers and amplifies regularities on their inputs that are statistically related across processors. It can be thought of as a kind of internal mutual observation that seeks agreement. The optimal state specified by the Coherent Infomax objective is computationally tractable only in simple situations, however. Therefore, in more complex situations it is a direction of travel, not a terminus. There are endless possibilities for specialized sensors, effectors, and connection architectures. This is consistent with the great diversity that is seen in the findings of neuroscience. The hypothesis being examined here is that the probabilistic inference underlying all that diversity is subject to a single, unified, coherent logic, or computational theory.

4. JAYNES'S PROBABILITY THEORY

Edwin T. Jaynes (1922-1998) was a physicist who worked on quantum electrodynamics, statistical mechanics, information theory, and probability theory mostly in Washington University, St. Louis, but also in Stanford, Berkeley, Princeton, and MIT in the USA, and at Cambridge in the UK. His arguments for and developments of probability as a measure of uncertainty, rather than as the relative frequency of an outcome in the "long-run", remain highly influential in physics, mathematics, engineering, and machine-learning. Though a few neurobiologists have used Jaynes's ideas [24,25], they are not yet widely known in either biology or psychology. Of his many publications, those most used in preparing this outline are: "Probability theory as logic" [26]; "Where do we stand on maximum entropy? [27]; "Information theory and statistical mechanics" [28], and the posthumous book, edited by G. Larry Bretthorst, "Probability Theory: The Logic of Science" [1]. Specific references to particular writings will be given below only where necessary.

His central contribution to probability theory was an in-depth study of its use to quantify uncertainty. This rejected the frequentist definitions that had been dominant in statistics for many decades. Such a change in the definition of "probability' may seem unimportant, but it has major consequences, both conceptually and in real applications. Jaynes defines probability as quantifying the uncertainty of inferences drawn from given conditions. It is therefore often referred to as epistemic. By classical frequentist definitions probability quantifies properties of the observed world, i.e., it is ontological. Frequentist definitions apply only to populations of observations, but epistemic probabilities apply also to individual cases. Furthermore, epistemic probabilities allow inferences to use all relevant information, whereas frequentist conceptions use only relative frequencies. Jaynes's logic assumes that scientific inferences should be treated as working hypotheses that are continually improved by new observations. His theory shows how to update estimates of the plausibility of hypotheses by appropriately weighting new observations and priors according to the strength of the evidence on which they are based. Unjustified bias in estimating that weighting is minimized by using maximum entropy methods. These do not apply any law of physics, but are simply a way of making predictions without making arbitrary unjustified assumptions. In the simple case of dice throwing, for example, this allocates equal probability to all six possible outcomes, if no other information is available. Other information, of endless forms, can modify these probabilities, including inspection of the dice and the surface on which it is thrown, knowledge of the honesty of the owner, and the relative frequencies of previously observed outcomes. Frequentist conceptions of probability are thus treated as a special limited case of Jaynes's more general interpretation of probability.

Within Jaynes's theory nothing is assumed to happen by chance or "at random" in reality; instead, he argued that randomness is a slippery, undefined, and unverifiable notion [26]. As a physicist, he argued against the widely held Copenhagen interpretation of quantum mechanics, in which quantal uncertainty is assumed to be a property of the world. He insisted, instead, that that could never be proved, so that all we can say is that no deterministic rules governing quantal events have yet been discovered. Discussing Boltzmann's statistical mechanics, he notes that it is sometimes asked where the randomness necessary for statistical behavior comes from if the universe is at heart an orderly, deterministic place. Jaynes sees this as a non-problem because to him probability theory as logic easily explains what we see, as a consequence of physical laws even if they are deterministic. The theorems of statistical mechanics are then interpreted as inferences predicting macroscopic behavior as best we can when given only knowledge of macroscopic variables that are interpreted as averages over microstates. More generally, he makes no use of the notion of a "random variable", even though that is usually presented as fundamental to all statistics. He noted that use of this notion requires assumptions that are usually unnecessary and often implausible, such as ergodicity. Fortunately, resolution of this debate concerning ultimate determinism is not crucial here because adaptively self-organized systems depend upon what is in practice predictable, and, even in a deterministic world, that is severely constrained by incomplete information, deterministic chaos, or computational feasibility. Thus, probabilistic inference will usually be needed in practice because uncertainty is so common.

It is often asked "whose" uncertainties are quantified by using epistemic probabilities. When discussing Shannon's information entropy, Jaynes [1] explicitly concludes that it is the uncertainty of the designer of the communication system, i.e., the engineer. Self-organized systems do not have a designer, however, so here we need a different answer. When thinking about this problem we may be misled by phenomenological connotations of "uncertainty", but they can be avoided by further study of Jaynes's theory. It extends logic to show what inferences concerning the plausibility of given hypotheses can be validly drawn from given conditions. Deductive logic is the special case where inferred probabilities are 0 or 1. Putting it this way shows that probabilistic inference is relative to the given conditions, as is deduction. As that relativity is unavoidable, this shows that Jaynes's logic is as "objective" as deductive logic.

Though Jaynes played a leading role in initiating the "Bayesian" revival in statistics and beyond, and refers to his theory throughout as Bayesian, very little was actually contributed by Thomas Bayes himself. The terms "Bayes" and "Bayesian" are little more than custom without content; replacing them with "Jaynes" and "Jaynesian" would be both more accurate and more useful. It was Laplace whose writings pre-figured more of the conception for which Jaynes argued so passionately and extensively. Jaynes goes well beyond both. He formally derived the whole framework from a few elementary logical desiderata, the most fundamental being the requirement of consistency. He showed how probability theory applies to non-equilibrium states. He showed how thermodynamic entropy and information-theory entropy (uncertainty) can be interpreted as the same concept, and not merely as sharing a mathematical expression. He emphasised the importance of distinguishing epistemic from frequentist definitions, which Bayes did not. He showed how classical statistical methods and frequentist probability definitions are essentially special cases of his methods and definitions. He established Maximum Entropy methods as the best way to set priors for things unknown. These methods are now used for data analysis and prediction in a wide range of applications.

For all of these contributions he deserves widespread acknowledgement and attention.

5. RELATIONS BETWEEN JAYNES'S PROBABILITY THEORY AND ADAPTIVELY SELF-ORGANIZED COMPLEXITY

Jaynes presents his probability theory as the logic of science [1]. Here it is considered as the logic of life. There is a deep ambiguity within Jaynes's theory. It is primarily concerned with the logic by which scientists should make explicit inferences. The logic proposed is so general and unified, however, that, if correct, it underlies all valid inference, whether made explicitly by scientists or implicitly by anything. If living organisms adapt to their environment by implicitly making inferences about it, then it is possible that probability theory could provide a conceptual framework within which to understand how they do so. I assume that any such contribution would be at the level of computational theory. i.e., it would clarify the logic underlying all inference, while leaving open issues concerning the particular strategies (representations and algorithms) and physical mechanisms by which requirements of that logic are met or approximated.

Jaynes [27] related his theory to biology by discussing potential uses of his probability theory to infer organisms' macroscopic properties from knowledge of their molecular structure. He suggested that this might be possible in some simple cases within a few decades. The issue here is not with what inferences science can draw about biological systems, however, but with principles of inference implicit in biological activity. We must therefore take a "first-person" view of information and inference from the perspective of the system itself [24,25]. This requires us to make inferences about inferential systems.

Jaynes himself was very inconsistent on this issue. His logic is often presented as prescriptive, showing how inferences should be drawn, and as requiring a scientist, or someone else, to specify the issues to be studied, to propose hypotheses, to provide the priors, and select the relevant data. He sometimes explicitly denied that the theory was descriptive of the way in which people or other organisms do in fact infer. In 1958 he submitted a paper to a physics journal using the title "How does the brain do plausible reasoning?" In response to its rejection he said that the theory presented was not meant seriously as a description of how real brains actually do perform plausible reasoning, but as showing how an idealized "robot" brain should reason. Elsewhere, he argued that familiar problems of everyday life are typically so much more complicated than scientific problems, and depend on so many unknown and uncontrolled factors, that a full Bayesian analysis is out of the question in practice [1]. In replying to criticisms of the utility of Bayesian methods, he says "The first critic objected to Bayesian methods on the grounds that they do not tell us how to create hypotheses (although neither do any other methods). This is like criticizing a computer because it will not push its own buttons; of course, it is up to us to tell Bayesian theory which problem we want it to solve. Would anybody want it otherwise?" [26].

In contrast to such denials of probability theory as descriptive of implicit inference in biological systems, Jaynes more often argues for its applicability to inference of any kind. He showed in detail how his probability theory can be applied to irreversible non-equilibrium processes as well as to equilibrium processes [29], and that suggests potential relevance to self-organized complexity. His belief in the wider underlying generality of his proposed inferential principles is also made clear in his first paper on information theory and statistical mechanics [28], where he concludes that Shannon entropy, used as a measure of uncertainty, becomes the primitive concept with which we work, more fundamental even than energy.

He later concluded that “seeing is inference from incomplete information”, and that probability theory is “telling us something about the way our own minds operate” when we unconsciously form intuitive judgments [1]. He explicitly argued that any reasoning which conflicts with Bayesian principles would place a creature at a decided survival disadvantage, so evolution by natural selection would automatically produce brains which reason in the Bayesian format. In reply to the claim that human reasoning is often not Bayesian, Jaynes says: “One disadvantage of having a little intelligence is that one can invent myths out of his own imagination, and come to believe them. Wild animals, lacking imagination, almost never do disastrously stupid things out of false perceptions of the world about them. But humans create artificial disasters for themselves when their ideology makes them unable to perceive where their own self-interest lies.” [26]. This implies that Bayesian inference is common across species, but in humans operates under special limitations that are a consequence of their own creative imagination. He also suggests that conscious human reasoning is sometimes misled by “false indoctrination”.

Science has done for humans what nothing has done for any other species. Therefore, principles of Jaynes’s probability theory that distil the essence of inference in general cannot also distil any special essence unique to science. My working assumption is that science depends upon distinctively human cognitive capabilities that have somehow overcome constraints under which more widely embodied strategies, or algorithms, for inference operate. Probability theory, construed as the logic of science, requires explicit conscious hypothesis creation and testing by people such as scientists and engineers. In life more generally it must be self-organized, but how is that possible? In response to that fundamental mystery the following section reconsiders Coherent Infomax in the light of Jaynes’s logic.

6. RELATIONS BETWEEN JAYNES'S PROBABILITY THEORY AND COHERENT INFOMAX

Coherent Infomax was originally proposed as a multi-purpose algorithm implemented in mammalian neocortex [13]. In short, its task was to discover predictive relationships latent within rich datasets. Encouraged by the much greater generality of Jaynes's probability theory, and by the broad conception of biological inference in Friston's "free energy" theory [10,30], our working hypothesis is now that the relevance of Coherent Infomax extends well beyond neocortex. This broader potential is clearly implied by its formulation in terms of Shannon entropy. So, can it be combined with Jaynes's unified theory of probability as extended logic to provide a unified conception of brain function?

6.1. Challenges Faced by Theories of Self-Organized Inference in Neural Systems

Creation of such a unified theory faces difficult challenges, however. First, it must be shown how inference can be self-organized. Second, though the logical desiderata from which Jaynes begins seem simple, they are councils of unattainable perfection at the system level. We cannot guarantee that all our beliefs are consistent, and we can rarely be sure that we use all relevant knowledge. Furthermore, it could be argued that some inconsistencies should be tolerated, or even welcomed. Therefore, it may be better to treat Jaynes's desiderata at the system level as goals to be approximated, rather than as absolute requirements. Third, there is the difficulty well-known as the "curse-of-dimensionality" [31]. Most raw sensory data occurs within a space of such high dimensionality that individual events are distributed very sparsely throughout it, with most locations within the space being empty, and few with more than one prior sample. This is true for all realistic lifetimes, whether of individual or species, because the number of locations grows exponentially with dimensionality. Therefore, prior events within the raw sensory space taken as a whole cannot serve directly as a basis for inference. As there are enduring regularities within the real world, however, most events actually occur on or near manifolds of much lower dimension. Discovery of such manifolds is therefore a crucial problem to be solved. Fourth, inferential tractability decreases exponentially with the number of constraints to be consistently satisfied. How can that difficulty be alleviated? Fifth, another major difficulty arises from the ubiquity of ambiguity. This is best resolved by using predictions arising from the context, but how can that be done while avoiding self-fulfilling prophecies? Finally, any major transitions through which inferential capabilities have evolved need to be identified.

The following discussion outlines ways in which Coherent Infomax responds to these challenges, emphasizing the distinction between driving and contextual interactions. It then examines ways in which this distinction may be related to Jaynes's probability theory. Limitations in the extent to which Coherent Infomax explains higher cognitive functions will then be mentioned, together with a discussion of the need for a more differentiated account of major transitions in cognitive evolution.

6.2. How Coherent Infomax Responds to These Challenges

The Infomax component of Coherent Infomax formalizes a widely accepted principle for the self-organisation of efficient coding in neural systems. It was originally called the "reduction of redundancy" by Horace Barlow [32], then formalised and called Infomax by Ralph Linsker [33]. It maximizes the mutual information between input and output of a local processor under the constraint of substantial data compression, so it greatly eases the curse-of-dimensionality. Infomax is formulated using Shannon entropy, and increases information transmission by maximizing the variance of outputs given the actual inputs received. It can therefore be directly related to Maximum Entropy methods in the case of Gaussian noise [33].

Useful inference requires more than efficient coding, however. Sensory systems can provide so much information that the more difficult problem is separating the crucial variables from the "noise". When there is far too much information for any given purpose, more is required than compression of all the available information into a smaller amount of data. Coherent Infomax suggests a way of specifying what information to transmit; its objective is to discover and amplify those variables that are predictively related to the particular context in which they occur. It therefore provides a way of specifying what information is important. Elementary variables discovered to be crucial to survival by adaptive evolution can then be specified directly rather than having to be discovered through further learning. Coherent Infomax can then discover and amplify new variables defined upon richer data-sets that predict the crucial variables. Once evolved because of their contribution to survival, however, mechanisms implementing Coherent Infomax can be used to discover and create statistical structure that may have little or no relevance to survival. Similarly, in a practical application of Coherent Infomax, the engineer could directly specify any relevant variables that are known, and then use Coherent Infomax to find new variables that are predictively related to them. In addition, however, he could also use it for open, exploratory, data mining.

An unavoidable consequence of the curse-of-dimensionality is that large amounts of data must be divided into subsets that are small enough to make learning feasible. If they were processed independently, however, then relations between the subsets would be unobservable. Success in finding the relevant manifolds would then be completely dependent upon the original division into subsets, but that is unlikely to be adequate unless the manifolds were already known. Coherent Infomax responds to this dilemma by dividing data at each level of an interpretive hierarchy into many small subsets, and searching for variables defined on them that are predictively related across subsets. This strategy allows for endlessly many ways in which the data can be divided into subsets, and linked across subsets.

Grouping large datasets into smaller subsets can also make inference more tractable by limiting the number of constraints within which consistency is sought. Within many real situations, with large knowledge bases, the best that can be done is to maximise the local consistencies and minimize the local inconsistencies [34]. When performed dynamically within perception this is known as Gestalt grouping, and psychophysical studies have discovered many criteria on which it is based in biological perceptual systems. The importance and difficulty of these grouping processes is clearly demonstrated by the existence of a whole sub-discipline within machine perception devoted to the task of "perceptual organization". The information required for such dynamic grouping overlaps greatly with that required for contextual disambiguation, and much evidence suggests that within neural systems there is overlap between the mechanisms involved; furthermore, those mechanisms can be interpreted as implementing Coherent Infomax [14].

Contextual disambiguation is central to the Coherent Infomax strategy. Because the data to be interpreted by local processors within each level of the hierarchy arises only from a subset of the data, it will typically be compatible with a range of possible interpretations. Coherent choices across the system as a whole can be facilitated by amplifying those local choices that are most likely within the context of the activity of other processors. These contextual predictions must not by themselves be sufficient to drive local processor activity, however, because, if they were, self-fulfilling prophecy would remove the ability to learn about the real world. This is formalized within Coherent Infomax as the minimization of the conditional mutual information between outputs and contextual inputs given the driving inputs. Information specifically about the context is therefore not transmitted, thus ensuring that it does not corrupt the information transmitted specifically about the driving inputs. In neural systems this is implemented by using the contextual inputs to control the gain of the response to the driving inputs, and several synaptic and local circuit mechanisms for doing this have been discovered [13,14,15,35].

6.3. Can the Asymmetry Between the Effects of Contextual and Driving Inputs Be Related to Jaynes's Probability Theory?

The theory of Coherent Infomax is based on a fundamental asymmetry between the effects of contextual and driving inputs. In short, contextual inputs are neither necessary nor sufficient to produce an output; driving inputs are both necessary and sufficient. Can this asymmetry be related to Jaynes's probability theory? Prima facie, the most obvious possibility is that in Bayesian inference priors provide context-sensitivity. That is what I had long assumed, and it may seem obvious to many that context must operate via the prior. However, Jim Kay recently showed that not to be so [6]. In Coherent Infomax context contributes to the posterior via the likelihood, not via the prior. In Bayesian inference posteriors are proportional to the product of priors and likelihoods. There is no essential asymmetry between them. Strong priors can outweigh weak likelihoods, and vice versa. In its standard form, Bayesian inference is simply a way of accumulating data, and the distinction between earlier and later data is essentially arbitrary [1]. Priors are neither necessarily nor usually context-sensitive. What Jim Kay's new perspective shows is that, in our approach, driving inputs are used to compute prior probability distributions, and that context contributes to the computation of the likelihood of the data to be interpreted given the "hypothesis" whose probability is to be updated [6]. This is therefore equivalent to using context as part of the generative model in context-sensitive Bayesian techniques as mentioned below.

Bayesian inference does seem to imply an asymmetry in that the data is a "given" that is assumed to be true. Data is also "fixed" in the sense that probability distributions are computed for likelihoods over varied model parameters for the given data. This asymmetry is less relevant here than at first appears, however, because the hypotheses whose probabilities are estimated by Bayesian inference concern unknown things. The objects of inference are not the observations themselves, but uncertain things about which the data provides evidence, such as the parameters in a model of the underlying processes that generate the data, or future data yet to be observed. Priors can provide evidence about those parameters that is as strong as, or stronger than, the current data. Consider a gambler who uses some system or inside knowledge to bet on a horse that loses. He may nevertheless be justified in making further bets on the same grounds if that strategy pays off over many races. No single outcome has a privileged status in determining that.

An asymmetry equivalent to that emphasized by Coherent Infomax does occur in some uses of maximum entropy and Bayesian methods, however. In machine learning, for example, a probability distribution of possible translations of a new occurrence of a familiar word in a body of text may be estimated from a sample of prior translations. Conditional maximum entropy methods do this using context [36]. The item to be translated must have previously occurred at least once previously, and preferably many times, so it should be defined on a space of low dimensionality. The particular combination of item and context may be entirely novel, however; so it can occur within a space of much higher dimensionality. Presence of the item to be translated is both necessary and sufficient to produce an output distribution; the context is neither necessary nor sufficient. This is equivalent to the effects of context in Coherent Infomax. Context-sensitivity has also been added to Bayesian techniques in some other applications. In bioinformatic data mining, for example, context-sensitivity was found to make dramatic improvements in both the precision and the sensitivity of predictions [37]. Context-sensitive Bayesian techniques have also been applied to the task of correcting spelling errors that result in valid words, but where the context indicates that some other word was intended [38]. The role of context in all these applications is essentially equivalent to that in Coherent Infomax, so their practical success greatly encourages our study of its theoretical significance.

6.4. What Are the Major Transitions in the Evolution of Inferential Capabilities?

When showing how learning can proceed without necessitating storage of all the details of past experience, Jaynes [1] distinguished between conscious and subconscious inference. He suggests that subconscious processes continually update a density function for each probability to be estimated. This function has a broad distribution when the prior evidence is weak. It has a single high peak when prior evidence for the stored probability is strong. The former, but not the latter, is easily changed by new evidence. He suggests that in humans the subconscious system (which is similar to what psychologists call "procedural memory") may more reliably reflect the inferential principles that he proposes than does the conscious system. In contrast to that suggestion, advances in the cognitive and neurosciences support the view that Bayesian principles play a leading role in both conscious human reasoning [39], and unconscious inference [11,40]. Though that work still has far to go [41], and has so far made little use of Jaynes's advances on Bayes, it shows that, as he thought, his preliminary explorations of these issues, were early steps into a large new territory.

A crucial issue within that territory concerns the possibility of major transitions in the evolution of inferential capabilities. Szmathàry and Maynard Smith [42] identified seven major transitions in the evolution of life, such as the transition from asexual to sexual reproduction. Only one of those concerned cognition, i.e., the transition to language. Major transitions in the evolution of inferential capabilities prior to language are also possible, however, and it is crucial to determine whether this is so because empirical studies of inferential capabilities will be misinterpreted if they are assumed to reflect a single strategy, when instead they reflect a mixture of strategies, either across or within species.

Probability theory could contribute to this issue by proposing various possible inferential strategies. For example, these could range from those with requirements that are simple to meet but with severely limited capacities, through intermediate stages of development, to those having more demanding requirements but with enhanced capabilities. Some possible transitions are as follows: from predictions only of things that are directly observable to estimates of things not directly observable; from generative models averaged over various contexts to those that are context specific; from hypotheses determined by input data to those that are somehow more internally generated; from probabilistic inference to syntactic structure, and, finally, from hypothesis testing to pure hypothesizing freed from testing. Within stages marked by such transitions there would still be much to be done by gradual evolutionary processes. For example, context-sensitive computations can make astronomical demands on computational resources, so they are only useful if appropriate constraints are placed on the sources and size of contextual input, as already shown for its use in natural language processing [38]. Thus, even given the ability to use contextual information, the search for useful sources of contextual input could still be a lengthy process, even on an evolutionary timescale, and produce much diversity.

Transition from non-conscious to conscious strategies was not included in the list of possible transitions just given for the simple reason that all of those mentioned were explicitly related to probabilistic inference. It is not clear how that can be done for consciousness. Though it is not necessary for inference, the possibility that various aspects of consciousness are associated with one or more of the possible transitions listed may be an important issue for further research.

These speculations go far beyond the current theory of Coherent Infomax. Limitations of the theory are discussed elsewhere [43], but one of particular relevance here is that the theory does not even account for intentional representation [13]. It neither assumes nor explains the existence of an interpreter who knows of, and distinguishes, both signs and what they signal. Our hope is that the theory will contribute to our understanding of such higher cognitive functions by showing what can be done without them, thus revealing both the foundations on which they build, and the constraints they overcome.

The tension between creation and discovery implicit in Jaynes's view of the human imagination is memorably expressed in the quote from Montaigne with which he begins his 1990 paper [26]: "Man is surely mad. He cannot make a worm; yet he makes Gods by the dozen." Man's ability to make a worm is now closer to reality than Montaigne thought, and, creativity underlies life in general; not only does it form and reform itself, but it also transforms the world in which it lives. Our susceptibility to delusion may therefore be a price we pay for the capabilities that enable us to create organized worlds of ever-increasing complexity [15,44], though that hypothesis needs careful formulation [45].

7. DOES SELF-ORGANIZED INFERENCE IN LIVING THINGS HAVE AN OBJECTIVE?

Several theories of brain function derive system properties from a formally specified objective function whose value tends to change in one direction over time as the system evolves, in both the short-term and the long-term. Coherent Infomax is one of them. Though Jaynes proposed an underlying logic for plausible inference, he did not specify inferential objectives, which were assumed to be supplied by the scientist, engineer, or whoever else is making the inferences. Coherent Infomax adds to Jaynes's logic by proposing a formal objective function, and by showing, in principle, how that objective can drive the dynamics of adaptively self-organized complex systems. This objective is Jaynesian in spirit, however, because it produces patterns of activity in which the mutual information shared by elements is high, even though it also increases their joint information. This is "Jaynesian" because it increases the amount of information from which useful inferences can be drawn.

The theory of Coherent Infomax was developed independently of any particular definition of probability, but Fiorillo [46] argues strongly that the use of a unified Jaynesian definition would enable neuroscience advance beyond the state currently reached by using Bayesian theories without acknowledging any distinction between frequentist and epistemic probabilities. From that perspective, he argues that the computational goal of neural systems is to infer or predict states of the world for the purpose of deciding which motor outputs to select. Quantitatively he describes this as minimizing uncertainty about biologically relevant aspects of the world, or, equivalently, as maximizing information about those aspects [46].

This goal is similar to that of Coherent Infomax, and in both Fiorillo's theory [24,46] and ours inference is seen as central to brain function, and Jaynes's ideas are seen as providing an insight into its underlying logic that is far more profound than the mere Bayesian inversion of conditional probabilities.

If it is not only entropy, but also self-organized complexity that increases over much of cosmic history, then Richard Dawkins' selfishness is not the only option for a scientifically based conception of long-term objectives. We can think of life at the ecological and species levels, not as "evolved to reproduce", but as "reproducing to evolve"; i.e., in the direction of the formally specified objective. One answer to the question "Why does it do that?" is then simply "Because it can". From that perspective we can think of our own individual efforts as directed, not merely towards survival, but as directed towards whatever organized complexities we choose to create.

ACKNOWLEDGMENTS

Jim Kay formalized the Coherent Infomax theory. Christopher Fiorillo convinced me that Jaynes's probability theory has much to offer. My efforts on this paper show how much I value their thoughts. They also provided useful comments on an earlier version, as did Ron Cottam, Bob Doyle, Mike Spratling, and an anonymous referee. I am grateful to Gordana Dodig-Crnkovic, Editor of this Special Issue, for encouraging this work.

REFERENCES

1. Jaynes, E.T. Probability Theory: The Logic of Science; Bretthorst, G.L., Ed.; Cambridge University Press: Cambridge, UK, 2003.

2. Engel, C.; Singer, W. Better than Conscious? MIT Press: Cambridge, MA, USA, 2008.

3. Phillips, W.A.; Kay, J.; Smyth, D. The discovery of structure by multi-stream networks of local processors with contextual guidance. Netw. Comput. Neural Syst. **1995**, 6, 225–246.

4. Kay, J.; Floreano, D.; Phillips, W.A. Contextually guided unsupervised learning using local multivariate binary processors. Neural Network. **1998**, 11, 117–140.

5. Kay, J.; Phillips, W.A. Activation functions, computational goals and learning rules for local processors with contextual guidance. Neural Comput. **1997**, 9, 895–910.

6. Kay, J.; Phillips, W.A. Coherent Infomax as a computational goal for neural systems. Bull. Math. Biol. **2011**, 73, 344–372.

7. Sporns, O. Complexity. Scholarpedia **2007**, 2, 1623.

8. von der Malsburg, C.; Singer, W. Principles of Cortical Network Organization. In Neurobiology of Neocortex; Rakic, P., Singer, W., Eds.; John Wiley & Sons: Chichester, UK, 1988.

9. Rolston, H., III. Three Big Bangs: Matter-Energy, Life, Mind; Columbia University Press: New York, NY, USA, 2010.

10. Phillips, W.A.; von der Malsburg, C.; Singer, W. Dynamic Coordination in Brain and Mind. In Dynamic Coordination in the Brain: From Neurons to Mind; von der Malsburg, C., Phillips, W.A., Singer, W., Eds.; MIT Press: Cambridge, MA, USA, 2010.

11. Friston, K.J. The free-energy principle: A unified brain theory? Nat. Rev. Neurosci. **2010**, 11, 127–138.

12. Finger, S. Origins of Neuroscience; Oxford University Press: New York, NY, USA, 1994.

13. Phillips, W.A.; Singer, W. In search of common foundations for cortical computation. Behav. Brain Sci. **1997**, 20, 657–722.

14. Dynamic Coordination in the Brain: From Neurons to Mind; von der Malsburg, C., Phillips, W.A., Singer, W., Eds.; MIT Press: Cambridge, MA, USA, 2010.

15. Phillips, W.A.; Silverstein, S.M. Convergence of biological and psychological perspectives on cognitive coordination in schizophrenia. Behav. Brain Sci. **2003**, 26, 65–138.

16. Becker, S.; Hinton, G.E. A self-organizing neural network that discovers surfaces in random-dot stereograms. Nature **1992**, 335, 161–163.

17. Creutzig, F.; Globerson, A.; Tishby, N. Past-future information bottleneck in dynamical systems. Phys. Rev. **2009**, 79, 041925.

18. Kelso, J.A.S. Dynamic Patterns: The Self-Organization of Brain and Behavior; MIT Press: Cambridge, MA, USA, 1995.

19. Körding, K.P.; König, P. Learning with two sites of synaptic integration. Netw. Comput. Neural Syst. **2000**, 11, 1–15.

20. Salinas, E.; Sejnowski, T.J. Gain modulation in the central nervous system: Where behavior, neurophysiology, and computation meet. Neuroscientist **2001**, 7, 430–440.

21. Spratling, M.W. Predictive-coding as a model of biased competition in visual attention. Vis. Res. **2008**, 48, 1391–1408.

22. Spratling, M.W.; Johnson, M.H. A feedback model of perceptual learning and categorisation. Vis. Cogn. **2006**, 13, 129–165.

23. Tononi, G.; Sporns, O.; Edelman, G.M. Complexity and coherency: Integrating information in the brain. Trends Cogn. Sci. **1998**, 2, 474–484.

24. Fiorillo, C.D. Towards a general theory of neural computation based on prediction by single neurons. PLoS One **2008**, 3, e3298.

25. Fiorillo, C.D. A neurocentric approach to Bayesian inference. Nat. Rev. Neurosci. **2010**, 11, 605.

26. Jaynes, E.T. Probability Theory as Logic. In Maximum Entropy and Bayesian Methods; Fougere, P.H., Ed.; Kluwer Academic Publishers: Dordrecht, The Netherlands, 1990.

27. Jaynes, E.T. Where do We Stand on Maximum Entropy? In The Maximum Entropy Formalism; Levine, R.D., Tribus, M., Eds.; MIT Press: Cambridge, MA, USA, 1979.

28. Jaynes, E.T. Information theory and statistical mechanics. Phys. Rev. **1957**, 106, 120–130.

29. Jaynes, E.T. Information theory and statistical mechanics. II. Phys. Rev. **1957**, 108, 171–191.

30. Friston, K.J.; Stephan, K.E. Free-energy and the brain. Synthese **2007**, 159, 417–458.

31. Bellman, R.E. Adaptive Control Processes; Princeton University Press: Princeton, NJ, USA, 1961.

32. Barlow, H.B. Possible Principles Underlying the Transformation of Sensory Messages. In Sensory Communication; Rosenblith, W.A., Ed.; MIT Press: Cambridge, MA, USA, 1961.

33. Linsker, R. Self-organization in a perceptual network. Computer **1988**, 21, 105–107.

34. Dechter, R. From local to global consistency. Artif. Intell. **1992**, 55, 87–107.

35. Silver, R.A. Neuronal arithmetic. Nat. Rev. Neurosci. **2010**, 11, 474–489.

36. Berger, A.L.; Della Pietra, S.A.; Della Pietra, V.J. A maximum entropy approach to natural language processing. Comput. Linguist. **1996**, 22, 39–71.

37. Myers, C.L.; Troyanskaya, O.G. Context-sensitive data integration and prediction of biological networks. Bioinformatics **2007**, 23, 2322–2330.

38. Ginter, F.; Boberg, J.; Jarvinen, J.; Salakoski, T. New techniques for disambiguation in natural language and their application to biological text. J. Mach. Learn. Res. **2004**, 5, 605–621.

39. Oaksford, M.; Chater, N. Bayesian rationality: The probabilistic approach to human reasoning. Behav. Brain Sci. **2009**, 32, 69–120.

40. Knill, D.C.; Pouget, A. The Bayesian Brain: The role of uncertainty in neural coding and computation. Trends Neurosci. **2004**, 27, 712–719.

41. Jones, M.; Love, B.C. Bayesian fundamentalism or enlightenment? On the explanatory status and theoretical contributions of Bayesian models to cognition. Behav. Brain Sci. **2011**, 34, 169–188.

42. Szmatháry, E.; Smith, J.M. The major evolutionary transitions. Nature **1995**, 374, 227–232.

43. Phillips, W.A. How do Neural Systems Use Probabilistic Inference that is Context-Sensitive to Create and Preserve Organized Complexity? In Integral Biomathics: Tracing the Road to Reality; Simeonov, P.L., Smith, L.S., Ehresmann, A.C., Eds.; Springer: Berlin, Germany, 2011.

44. Berhendt, R.-P.; Young, C. Hallucinations in schizophrenia, sensory impairment, and brain disease: A unifying model. Behav. Brain Sci. **2004**, 27, 771–830.

45. Phillips, W.A. Belief in the primacy of fantasy is misleading and unnecessary. Behav. Brain Sci. **2004**, 27, 802–803.

46. Fiorillo, C.D. Beyond Bayes: On the need for a unified and Jaynesian definition of probability and information within neuroscience. Information **2012**. submitted for publication.

CHAPTER 4

An Essay on the Double Nature of the Probability

Paolo Rocchi[1,2], Leonida Gianfagna[1]

[1]IBM, Roma, Italy

[2]Luiss University, Roma, Italy

ABSTRACT

Classical statistics and Bayesian statistics refer to the frequentist and subjective theories of probability respectively. Von Mises and De Finetti, who authored those conceptualizations, provide interpretations of the probability that appear incompatible. This discrepancy raises ample debates and the foundations of the probability calculus emerge as a tricky, open issue so far. Instead of developing philosophical discussion, this research resorts to analytical and mathematical methods. We present two theorems that sustain the validity of both the frequentist and the subjective views on the probability. Secondly we show how the double facets of the probability turn out to be consistent within the present logical frame.

KEYWORDS

Theory of probability; Subjectivism; Frequentism; Theorem of large numbers; Probability dualism

1. INTRODUCTION

When Hilbert prepared the list of the most significant mathematical issues to tackle in the next future, he included the probability foundations in the group of 23 famous problems. Some years later, Kolmorogov established the axioms of the probability calculus that the vast majority of authors have accepted so far. The Kolmogorovian axioms of non-negativity, normalization and finite additivity provide the rigorous base for calculations; they solve the issue in the terms posed by Hilbert but do not entirely unravel the fundamentals of the probability calculus. The probability is a measure and the physical nature of this measure has not been explained in a definitive manner [1]. The relation between probability and experience looks to be rather controversial.

For centuries the law of large numbers elucidated the meaning of the probability. The relative frequency in a large sample of trials offered the indisputable evidence for the probability calculated on the paper but after the First World War a significant turning point occurred. The calculus of the probability was used increasingly in business management. Many problems that arose in this environment dealt with making decisions under uncertainty and the method based on the frequency was not applicable in these cases as it is usually impossible to accumulate wide experience to assess the probability of an economical event. The subjective interpretation of the probability was introduced in this environment and became increasingly popular. In the first time theorists laid charge of arbitrariness against the subjective theory and vehemently rejected this frame [2-4]. Later the Bayesian methods infiltrated professionals' environment and experimentalists learned to calculate the probability on the basis of acquired information [5].

However the pragmatic adoption of different statistical methods appears inconsistent due to the conceptual divergence emerging between the probability seen as a degree of belief and the probability as long-term relative frequency. Popper sums up this situation by the ensuing statement:

"In modern physics (...) we still lack a satisfactory, consistent definition of probability; or what amounts to much the same, we still lack a satisfactory axiomatic system for the calculus of probability, [in consequence] physicists make much use of probabilities without being able to say, consistently, what they mean by 'probability'" [6].

This paper is an attempt to show why and when the frequentist and subjective approaches can be considered compatible in point of logic. We mean to prove that the dualist view on probability turns out to be consistent using two theorems.

2. TWO THEOREMS

We calculate some properties of the probability in relation to the relative frequency under the following assumptions:

1) The relative frequency is the observed number q of successful events A for a finite sample of n trials

$$F(A_n) = q/n \tag{1}$$

We assume $F(A_n)$ as the counterpart of $P(A)$ in the real world.

2) We exclude that A is sure or impossible, and $X_1, X_2, X_3, \cdots, X_m$ are the possible results of A with finite expected value μ

$$\mu = \frac{X_1 + X_2 + X_3 + \cdots + X_m}{m} \tag{2}$$

In principle there are two extreme possibilities:

3) There are n trials with

$$n > 1 \tag{3}$$

The occurrences of A make a sequence of pair-wise independent, identically distributed random variables $X_1, X_2, X_3, \cdots, X_n$ with finite sample mean SM_n

$$SM_n = \frac{1}{n}\sum_{j=1}^{n} X_j \tag{4}$$

4) There is only one trial namely

$$n = 1 \tag{5}$$

2.1. Theorem of Large Numbers (Strong)

Suppose hypothesis (3) true, then we have that $\overline{X}_n$ converges almost surely to μ

$$P\left(\lim_{n\to\infty} SM_n = \mu\right) = 1 \tag{6}$$

Proof: There are several methods to prove this theorem. We suggest the recent proof by Terence Tao who reaches the result with rigor under the mathematical framework of measure theory [7].

Remarks: This theorem expresses the idea that as the number of trials of a random process increases, the percentage difference between the actual and the expected values goes to zero. It follows from this theorem that the relative frequency of success in a series of Bernoulli trials will converge to the theoretical probability

$$F(A_n) \to P(A) \quad \text{when } n \to \infty \tag{7}$$

The relative frequency $F(A_n)$ after n trials will almost surely converge to the probability $P(A)$ as n approaches infinity.

2.2. Theorem of a Single Number

Let (4) true, the relative frequency of the event is not equal to the probability of A

$$F(A_1) \neq P(A) \tag{8}$$

Proof: The random event A either occur or does not occur in a sole trial, the following values are allowed for the frequency

$$F(A_1) = 1/1 = 1 \tag{9}$$

$$F(A_1) = 0/1 = 0 \tag{10}$$

We have excluded that A is sure or impossible, thus (8) and (9) mismatch with $P(A)$ that is decimal.

3. ANALYTICAL APPROACH

Speaking in general, the abstract determination of the parameter $M(A)$ that refers to the argument A is neutral, namely pure mathematicians have no concern for the material qualities of M and of A; instead those tangible properties are essential for experimentalists who particularize the significance of $M(A)$ determined by A. As an example take the real function $y=f(x)$ that has immutable mathematical properties in the environments A, B and C beyond any doubt. Now suppose that the variables x_A, x_B and x_C are heterogeneous in physical terms, than the qualities of y_A, y_B and y_C cannot be compared. Even if one adopts identical algorithms to obtain y_A, y_B and y_C, the concrete features of y_A, y_B and y_C do not show any analogy.

Kolmogorov chooses the subset A belonging to the set space Ω as the flexible and appropriate model for the argument of the probability $P(A)$. Pure mathematicians usually assume that A is the known term of a problem and do not call into question the subset A. This behavior works perfectly until one argues on the theoretical plane but things go differently in applications. We have discussed a set of misleading and paradoxes depending on the poor analysis of the probability argument [8,9]. In this paper we dissect the physical characteristics of the probability through its argument.

Hypotheses (3) and (5) deal with the occurrences of a random event; and the subsets A_n and A_1 prove to have two special states of being present in the world. No doubt $P(A_1)$ and $P(A_n)$ are equal as numbers but A_1 and A_n turn out to be distinct arguments in Section 2. The theorems just demonstrated show that the probability has different applied properties depending on the argument, and we rewrite (7) in more precise terms:

$$F(A_n) \to P(A_n) \quad \text{when } n \to \infty \tag{11}$$

And result (8) becomes:

$$F(A_1) \neq P(A_1) \tag{12}$$

The distinction between A_n and A_1 is not new. Ramsey, De Finetti and other subjectivists calculate the probability of the single occurrence A_1 in contrast with Von Mises who coined the term "collective" to emphasize the large number of occurrences pertaining to A_n. Various dualists argue upon these extreme possibilities such as Popper who develops the long-run propensity theory associated with repeatable conditions, and the single-case propensity theory.

It is necessary to note that philosophers, who argue on the probabilities of single and multiple events, are inclined to integrate the probability together with its argument. Generally speaking philosophers search for a comprehensive solution, they melt the probability with its argument and obtain a synthetic theoretical notion. Instead the present inquiry is analytical. We keep apart A_n from A_1 and later deduce the features of P from each argument. To exemplify statement (11) specifies the property of A_n instead one cannot tell the same for the argument A_1. Theorem 2.1 does not provide conclusions for the probability of a single trial, but solely for $P(A_n)$ in that $P(A_n)$ depends on A_n.

3.1. When the Number Is Large

Result (11) details that the relative frequency obtained after repeated trials in a Bernoulli series conforms to the probability.

Because of this substantial experiment one can conclude that $P(A_n)$ is a physical measurable quantity

$$P(A_n) \text{ is a real quantity in the world.} \tag{13}$$

The weak point of Theorem 2.1 is the limit conditions for the validation of $P(A_n)$ Solely infinite trials can corroborate the quantity $P(A_n)$ and nobody can bring forth this ideal experiment. If one does not reason in abstract, he can conclude that the ideal test does not place insurmountable obstacles. Experimentalists are aware that the higher is n and the better is the achieved validation of $P(A_n)$. The larger is the number of trials, the more the probability finds solid confirmation namely one has the rigorous method to validate the probability in the world out there.

Theorem of large numbers has long-term prehistory. Since immemorial time gamblers and laymen had keen interest in the corroboration of the values obtained through previsions, and they sensed that if an experiment is repeated a large number of times then the relative frequency with which an event occurs tends to the theoretical possibilities of that event.

Ample literature deals with the large number theorem and further comments seem unnecessary to us. Instead the single number theorem requires more accurate annotations.

3.2. When the Number Is the Unit

A premise turns out to be necessary to the discussion of Inequality (12).

Any hypothesis must be supported by appropriate validation in sciences. Theories and assertions are to be confirmed by evidences and not just by mental speculation [10]. Scientists search, explore, and gather the evidences that can sustain or deny whatsoever opinion. Popper highlights how a sole experiment can disprove a theoretical frame. Many candidate interpretations have been proposed by theoretical physicists during the centuries that have not been confirmed experimentally.

Test is supreme and a quantity that cannot be verified through trials is not extant. For examples some geologists suppose that earthquakes are able to emit premonitory signals. Forewarning signs could save several lives, but accurate tests do not bring indisputable evidences thus the scientific community concludes that premonitory signals do not exist.

Equation (12) holds that the frequency $F(A_1)$ systematically mismatches with $P(A_1)$ and this entails that one cannot validate the probability of A_1 in the physical reality. It is not a question of operational obstacles, errors of instruments, inaccuracy or other forms of limitation. Statement (12) establishes that in general one cannot corroborate the probability $P(A_1)$ by means of tests thus Theorem 2.2 leads to the following conclusion:

$$P(A_1) \text{ is a quantity that does not exist in the world.} \tag{14}$$

The probability of a single event is a number written on the paper which none can control in the real world.

This conclusion turns out to be dramatic as statistics play crucial role in sciences and modern economies. Statistics offers the necessary tools to investigate and understand complex phenomena. Moreover people have strong concern in unique events whose outcomes sometimes are essential for individuals to survive. In order to not throw away $P(A_1)$ authors established to bypass (14) in the following way.

The single-case probability has symbolic value, it is a sign and as such carries out communication task. $P(A_1)$ is a piece of information that can influence the human soul and can sustain personal credence. Thus the probability $P(A_1)$ may be used to express the degree of belief of an individual about the outcome of an uncertain even. Theorists do not discard the probability $P(A_1)$ and ground the subjective theory of probability upon this original function assigned to $P(A_1)$.

In principle subjective probability could vary from person to person, and subjectivists answer back the accusation of arbitrariness. They do not deem "the degree of belief" as something measured by strength of feeling, but in terms of rational and coherent behavior. The "Dutch book" criterion establishes a guide for the agent's degrees of belief which have to satisfy synchronic coherence conditions. An agent evaluates the plausibility of a belief according to past experiences and to knowledge coming from various sources, and this criterion opens the way to the Bayesian methods [11].

Theorem 2.2 has significant antecedents.

For centuries the vast majority of philosophers shared the deterministic logic and rejected any form of probabilistic issues. This orientation emerged in the Western countries since the classical era. In fact thinkers founded out that the possible results of a single random event yield unrealistic conclusions. As an example, suppose to flip a coin, one forecasts two balanced possibilities, instead a single test exhibits head or tail. Indeterminist logic suggests two outcomes equally likely, conversely experience shows only one outcome. Philosophers deemed a logic topic not entirely determined as unreliable and in consequence of this severe judgment the apodictic reasoning dominated the scientific realm. This systematic orientation kept scientists from paying adequate attention to probabilistic problems, and statistical methods infiltrated scientific environments solely by the end of the nineteenth century [12].

One can apply the frequentist theory provided the set of trials is very large instead the Bayesian statistics works even for a sole trial. The subjective theory does not encounter the obstacles typical of the frequentist frame. De Finetti and others recognize the theorem of large number as a property of the single event that repeats several times and conclude that the subjective theory is universal [13].

In consequence of (13) and (14) we can answer in the following terms.

Theorem 2.1 holds that $P(A_1)$ is a real parameter which one can validate in the world; conversely theorem 2.2 entails that $P(A_1)$ translates a personal belief into a number. The indefinite series $P_1(A_1), P_2(A_1), P_3(A_1), P_4(A_1)\cdots$ mirrors personal credence upon the repeated event A_1 and cannot be compared to $P(A_n)$ which is an authentic measure controllable in the physical reality. Hence it is worth the ponderous verification of $P(A_n)$ since this is a physical measure; instead $P(A_1)$ is a symbol.

It is true that $P(A_1)$ is flexible to countless single situations, but the theorem of a single number proves that $P(A_1)$ does not have any counterpart in the physical reality.

4. THE NATURE OF THE PROBABILITY

Least we can answer from Theorems 2.1 and 2.2 whether the probability is a measure with physical meaning, or is not physical or even has a double nature.

Theorem 2.1 demonstrates that $P(A_n)$ have different properties respect to $P(A_1)$, namely the frequentist nature and the subjective nature of probability relay on two separated situations. Therefore (13) and (14) prove to be compatible in point of logic since $P(A_1)$ and $P(A_n)$ refer to distinct physical events. It may be said that $P(A)$ is a real, controlled measure when:

$$A = A_n \tag{15}$$

Instead $P(A)$ has a personal value when:

$$A = A_1 \tag{16}$$

It is reasonable to conclude that the probability is double in nature and professional practice sustains this conclusion as the classical statistics and the Bayesian statistics investigate different kinds of situations. They propose methods that refer to separated environments. To exemplify physicists who validate a general law adopt the Fisher statistics; instead when a physicist has to make decision on a single experiment usually he adopts the Bayesian methods.

5. CONCLUSIONS

The interpretation of the probability is considered a tricky issue since decades, and normally theorists tackle this broad argument from the philosophical stance. The philosophical approach yielded several intriguing contributions but did not provide the definitive solution so far. We fear philosophy is unable to bring forth the overwhelming proof upon the nature of the probability because of its all-embracing method of reasoning. Instead the present paper suggests the analytical approach which keeps apart the concept of probability from its argument and examines the influence of A_1 and A_n over P.

Nowadays some statisticians are firm adherents of one or other of the statistical schools, and endorse their adhesion to a school when they begin a project; in a second stage each expert uses the statistical method which relies on his opinion. However this faithful support for a school of thought may be considered an arbitrary act as relying on personal will.

Another accepted view is that each probabilistic theory has strengths and weaknesses and that one or the other may be preferred. This generic behavior appears disputable in many respects since one overlooks the conclusions expressed by Von Mises which clash against De Finetti's conceptualization.

The present frame implies that the use of a probability theory is not a matter of opinion and suggests a precise procedure to follow in the professional practice on the basis of incontrovertible facts. Firstly an expert should see whether he is dealing with only one occurrence or with a high number of occurrences. Secondly an expert should follow the appropriate method of calculus depending on the extension (15) or (16) of the subset A.

The present research is still in progress. This paper basically focuses on classical probability and we should extend this study to quantum probability in the next future.

REFERENCES

1. D. Gillies, "Philosophical Theories of Probability," Routledge, London, 2000.
2. D. G. Mayo, "Error and the Growth of Experimental Knowledge," University of Chicago Press, Chicago, 1996.
3. M. R. Forster, "How Do Simple Rules 'Fit to Reality' in a Complex World?" Minds and Machines, Vol. 9, 1999, pp. 543-564.
4. M. Denker, W. A. Woyczyński and B. Ycar, "Introductory Statistics and Random Phenomena: Uncertainty, Complexity and Chaotic Behavior in Engineering and Science," Springer-Verlag Gmbh, Boston, 1998.
5. G. D'Agostini, "Role and Meaning of Subjective Probability; Some Comments on Common Misconceptions," 20th International Workshop on Bayesian Inference and Maximum Entropy Methods in Science and Engineering, AIP Conference Proceedings, Vol. 568, 2001, 23 p.
6. K. R. Popper, "The Logic of Scientific Discovery," Routledge, New York, 2002.
7. T. Tao, "The Strong Law of Large Numbers," 2012. http://terrytao.wordpress.com/2008/06/18/the-strong-law-of-large-numbers/

8. P. Rocchi and L. Gianfagna, "Probabilistic Events and Physical Reality: A Complete Algebra of Probability," Physics Essays, Vol. 15, No. 3, 2002, pp. 331-118.

9. P. Rocchi, "The Structural Theory of Probability," Kluwer/ Plenum, New York, 2003.

10. B. Gower, "Scientific Method: A Historical and Philosophical Introduction," Taylor & Francis, London, 2007.

11. L. J. Savage, "The Foundations of Statistics," Courier Dover Publications, New York, 1972.

12. J. Earman, "Aspects of Determinism in Modern Physics," In: J. Butterfield and J. Earman, Eds., Philosophy of Physics, Part B, North Holland, 2007, pp. 1369-1434.

13. B. De Finetti, "Theory of Probability: A Critical Introductory Treatment," John Wiley & Sons, New York, 1975.

CHAPTER 5

Strong Law of Large Numbers under an Upper Probability

Xiaoyan Chen

Graduate Department of Financial Engineering, Ajou University, Suwon, South Korea

ABSTRACT

Strong law of large numbers is a fundamental theory in probability and statistics. When the measure tool is nonadditive, this law is very different from additive case. In 2010 Chen investigated the strong law of large numbers under upper probability V by assuming V is continuous. This assumption is very strong. Upper probabilities may not be continuous. In this paper we prove the strong law of large numbers for an upper probability without the continuity assumption whereby random variables are quasi-continuous and the upper probability is generated by a weakly compact family of probabilities on a complete and separable metric sample space.

KEYWORDS

Strong Law of Large Numbers; Upper Probability; Weakly Compact; Independence; Quasi-Continuous

1. INTRODUCTION

Strong law of large numbers under nonadditive probabilities is a much important theory in uncertainty theories and has more applications in statistics, risk measures, asset pricings and many other fields. In 1999 Marinacci [1] first investigated the strong law of large numbers for sequences of independent and identically distributed (IID for short) random variables $\{X_i\}_{i=1}^{\infty}$ relative to a capacity ν which is continuous and totally monotone and proved that under regularity condition the limit inferior and limit superior of $\frac{S_n}{n}$, where $S_n := \sum_{i=1}^{n} X_i$, lie between the two Choquet integrals $\int X_1 \mathrm{d}\nu$ (submean) and $-\int -X_1 \mathrm{d}\nu$ (supermean) induced by this capacity with probability 1 under ν (that is, quasi surely), and furthermore, if ν is null-additive, then that limit inferior attains the submean and the limit superior attains the supermean quasi surely, respectively. This is different from the law under probability measure P whereby under suitable conditions, such as for IID sequences, $\frac{S_n}{n}$ converges to the mathematical expectation of X_1 almost surely relative to P. In 2005 Maccheroni and Marinacci [2] extended the results of Marinacci [1] for a totally monotone capacity on Polish space whereby the bounded variables X_i are continuous or simple, or the capacity ν is continuous. But the conditions of these two articles on capacity are too strong and not easy to test. And generally capacities can not uniquely determine the (nonlinear) expectations relative to the capacities. Motivated by robust statistics and limit theories under sublinear expectations given by Peng in 2007, Chen [3] in 2010 investigated the strong law of large numbers for a pair of lower and upper probabilities (v,V) which are induced by a sublinear expectation $\mathbb{E}$ (see Peng (2012) [4])

whereby the sequence $\{X_i\}_{i=1}^{\infty}$ is IID under $\mathbb{E}$ (the independence is different from classical case and the one in Marinacci [1], see Peng (2012) [4]). He proved if $\mathbb{E}\left[|X_1|^{1+\alpha}\right]<\infty$ for some $\alpha>0$, then those limit inferior and limit superior lie between another submean $-\mathbb{E}[-X_1]$ and supermean $\mathbb{E}[X_1]$ which may not equal the ones given by Choquet integrals (see Chen, Wu and Li (2012) [5] for details). Furthermore, if we futher assume that V is continuous, then

$$V\left(\limsup_{n\to\infty}\frac{S_n}{n}=\mathbb{E}[X_1]\right)$$
$$=V\left(\liminf_{n\to\infty}\frac{S_n}{n}=-\mathbb{E}[-X_1]\right)=1 \qquad (1)$$

Hu (2012) [6] extends the results of Chen [3] to the sequence of non-identically distributed random variables for the same independence and continuity assumptions. Chen and Wu (2011) [7] extends Chen [3] to more weaker independence condition without identical distribution assumption, and proves if we further assume that V is continuous, for any subsequence $\{n_i\}_{i=1}^{\infty}$ of $\mathbb{N}$, $\{S_{n_i}-S_{n_{i-1}}\}_{i=1}^{\infty}$ are pairwise weakly independent under v, and there exist some constants $\alpha>0,\ n_0\geq 1,$ and $c_0>0$ such that

$$\sup_{i\geq 1}\hat{E}\left[\left|X_i\right|^{1+\alpha}\right]<\infty, \frac{\left|S_n\right|}{n}\leq c_0 \ln(n+1), \forall n\geq 1,$$

then (1) still holds.

We can see that for the strong law of large numbers (1) under an upper probability, there are two key conditions: well-defined independence and continuity of the upper probability V. The continuity assumption of V is based on the second Borel-Cantelli lemma to get $V(A_n, i.o.)=1$ for certain sequence of measurable events A_n. But as we know in general V is not continuous, since for nonclosed nonincreasing sequence of measurable events G_n, even if $G_n \downarrow G, V(G_n)\downarrow V(G)$ may not hold (see Xu and Zhang (2010) [8] for an example). Hence a natural question is: if V is not continuous, whether does (1) hold? In this paper we will give a confirmative answer. We assume a1) Ω is a complete and separable metric space, F is a σ -algebra of all Borel subsets of Ω, P is a nonempty subset of $\mathcal{M}$ which is a family of all probabilities on (Ω, F), and P is also weakly compact;

a2) For each $i\geq 1$, X_i is quasi-continuous, and

$$\sup_{i\geq 1}\mathbb{E}\left[\left|X_i\right|^2\right]<\infty;$$

a3) $\{X_n\}_{n=1}^{\infty}$ is independent sequence of random variables under $\mathbb{E}$ where $\mathbb{E}$ is a sublinear expectation corresponding to P. In this paper we successfully proved the strong law of large numbers under assumptions a1)-a3) without the continuity assumption of V by transforming that an event $A\in F$ occures with probability 1 under V (that is, $V(A)=1$) to the problem of its complementary event A^c, i.e., $v(A^c)=1-V(A)=0,$ where v is the conjugate lower probability of V and proving $v(B_n,i.o.)=0$ for some appropriate events B_n with $\limsup_{n\to\infty} B_n \supset A^c$ by using properties of V and v.

This paper is organized as follows. In Section 2 we give some basic concepts and useful lemmas. In Section 3 we mainly prove the strong law of large numbers without continuity assumption of upper probability V for IID and continuous sequences. Section 4 extends results of Section 3 and gets the law for non-identically distributed sequence. Section 5 gives an example.

2. PRELIMINARIES

Let Ω be a separable and complete metric space. F is a σ-algebra of all Borel subsets of Ω. We introduce an upper probability V by

$$V(A)=\sup_{Q\in P} Q(A), \forall A\in F,$$

where P is a family of probabilities on (Ω,F) and weakly compact. Thus its conjugate capacity (see Choquet (1954) [9]), i.e., lower probability is

$$v(A)=\inf_{Q\in P} Q(A)=1-V\left(A^c\right), \forall A\in F,$$

where A^c is the complementary set of A. From Huber and Strassen (1973) [10] V and v also satisfy the following properties.

Proposition 1.

1) $V(\Omega)=v(\Omega)=1, V(\varnothing)=v(\varnothing)=0.$

2) $V(A)\leq V(B), v(A)\leq v(B)$ if $A\subset B, A, B\in F.$

3) $V\left(\sum_{n=1}^{\infty}A_n\right)\leq \sum_{n=1}^{\infty}V(A_n), \forall A_n\in F, n\geq 1.$

4) $V(A)\geq v(A)$ for all $A\in F.$

5) lower-continuity of V for all sets in F: if $A_n\uparrow A$, $A_n\in F$, then $V(A_n)\uparrow V(A).$

6) upper-continuity of V for all closed sets: if $F_n\in F$ closed, $F_n\downarrow F$, then $V(F_n)\downarrow V(F).$

7) lower-continuity of v for all open sets: if $G_n\uparrow G, G_n\in F$ open, then $v(G_n)\uparrow v(G).$

8) upper-continuity of v for all sets: if $B_n\in F, B_n\downarrow B$, then $v(B_n)\downarrow v(B).$

Now we introduce an upper expectation $\mathbb{E}$ by V in the following

$$\mathbb{E}[X]=\sup_{Q\in P}E_Q[X]$$

for all $X\in F$ such that $\sup_{Q\in P}E_Q[X]<\infty$, E_Q is the linear expectation corresponding to $Q\in P$ such that $E_Q[I_A]=Q(A), \forall A\in F.$

Then $\mathbb{E}$ is a sublinear expectation (see Peng [4]) on (Ω, H), where H is a set of all real-valued random variables $X \in F$ such that $\mathbb{E}[X] < \infty$, that is, $\mathbb{E}$ satisfies that for all $X, Y \in H$ 1) Monotonicity: $\mathbb{E}[X] \leq \mathbb{E}[Y]$ if $X \leq Y$.

2) Constant preserving: $\mathbb{E}[c] = c, \forall c \in \mathbb{R}.$

3) Sub-additivity: $\mathbb{E}[X+Y] \leq \mathbb{E}[X] + \mathbb{E}[Y].$

4) Positive homogeneity: $\mathbb{E}[\lambda X] = \lambda \mathbb{E}[X], \forall \lambda \geq 0.$

$(\Omega, H, \mathbb{E})$ is called a sublinear expectation space in contrast with probability space. Given $X = (X_1, X_2, \cdots, X_n)$, we say $X \in H^n$, if $X_i \in H$ for all $i = 1, 2, \cdots, n.$ For $X \in H$, $\mathbb{E}[X]$ is called its supermean, whereas $-\mathbb{E}[-X]$ is called its submean. If $\mathbb{E}[X] \neq -\mathbb{E}[-X]$, then X is said to have mean uncertainty.

In the following we introduce some useful concepts (one can refer to Peng (2010) [4] for details).

Definition 2. An n-dimensional random vector $Y \in H^n$ is said to be independent from an m-dimensional random vector $X \in H^m$ under $\mathbb{E}$, if for all bounded Lipschitz continuous functions $\varphi \in C_{b,Lip}\left(\mathbb{R}^{m+n}\right)$, we have

$$\mathbb{E}[\varphi(X,Y)] = \mathbb{E}\left[\mathbb{E}[\varphi(x,Y)]_{x=X}\right],$$

where $\mathbb{E}[\varphi(x,Y)]_{x=X} \in H^m.$

Remark 3. In general Y being independent of X under $\mathbb{E}$ does not imply X being independent of Y. See Example 3.13 of Chapter I in Peng (2012) [4] as a counterexample.

Definition 4. A sequence $\{X_n\}_{n=1}^{\infty}$ on $(\Omega, H, \mathbb{E})$ is said to be a sequence of independent random variables under $\mathbb{E}$, if for any $i \geq 1$, X_{i+1} is independent of $(X_1, X_2, \cdots, X_i)$ under $\mathbb{E}$

Definition 5. A real random variable $X \in F$ is said to be quasi-continuous (q.c. for short) if for any $\varepsilon > 0$, there exists an open set O with $V(O) < \varepsilon$ such that X is continuous on O^c.

Lemma 6. (Denis-Hu-Peng (2011) [11] Theorem 2) For any $A \in F$,

$$V(A) = \sup\{V(K) : \text{compact } K \subset A\}.$$

Remark 7. Lemma 6 implies that for any $A \in F$,

$$v(A) = \inf\{v(G) : \text{open } G \supset A\}.$$

The following Borel-Cantelli lemma is obvious (the readers also can refer to Peng [4] or Chen [3]).

Lemma 8 (Borel-Cantelli Lemma). For any sequence of events $\{A_n\}_{n=1}^{\infty}$ in F, if $\sum_{n=1}^{\infty} V(A_n) < \infty$, then

$$V(A_n, i.o.) = V\left(\bigcap_{k=1}^{\infty} \bigcup_{n=k}^{\infty} A_n\right) = 0.$$

Lemma 9 (Hu [6] Theorem 3.1). Let $\{X_n\}_{n=1}^{\infty}$ be a sequence of independent random variables on

$(\Omega, H, \mathbb{E})$. We assume 1) For any $i \geq 1$, there exist real constants $\underline{\mu}_i \leq \overline{\mu}_i$,

such that $\mathbb{E}[X_i] = \overline{\mu}_i$ and $-\mathbb{E}[-X_i] = \underline{\mu}_i$;

2) There exist two real constants $\underline{\mu} \leq \overline{\mu}$ such that

$$\lim_{n\to\infty} \frac{1}{n} \sum_{i=1}^{n} |\underline{\mu}_i - \underline{\mu}| = 0, \lim_{n\to\infty} \frac{1}{n} \sum_{i=1}^{n} |\overline{\mu}_i - \overline{\mu}| = 0;$$

3) $\sup_{n\geq 1} \mathbb{E}\left[X_n^2\right] < \infty.$

Set $S_n = \sum_{i=1}^{n} X_i, \forall n \geq 1.$ Then for any continuous function φ with linear growth on $\mathbb{R}$, we have

$$\lim_{n\to\infty} \mathbb{E}\left[\varphi\left(\frac{S_n}{n}\right)\right] = \sup_{\underline{\mu} \leq y \leq \overline{\mu}} \varphi(y).$$

Lemma 10 (Hu [6] Theorem 3.2 (I)). Let $\{X_n\}_{n=1}^{\infty}$ satisfy all the conditions given in Lemma 9, then

$$V\left(\limsup_{n\to\infty} \frac{S_n}{n} > \overline{\mu}\right) = V\left(\liminf_{n\to\infty} \frac{S_n}{n} < \underline{\mu}\right) = 0.$$

3. STRONG LAW OF LARGE NUMBERS

Theorem 11. Let $\{X_n\}_{n=1}^{\infty}$ be an independent and continuous sequence under $\mathbb{E}$. We assume there exist two real constants $\underline{\mu} \leq \overline{\mu}$ such that $\mathbb{E}[X_i] = \overline{\mu}$,

$-\mathbb{E}[-X_i] = \underline{\mu}$ for all $i \geq 1$, and $\sup_{n\geq 1} \mathbb{E}\left[|X_n|^2\right] < \infty.$

Set $S_n = \sum_{i=1}^{n} X_i, \forall n \geq 1.$ Then

$$V\left(\limsup_{n\to\infty} \frac{S_n}{n} = \overline{\mu}\right) = 1, \tag{2}$$

$$V\left(\liminf_{n\to\infty} \frac{S_n}{n} = \underline{\mu}\right) = 1. \tag{3}$$

Proof. It is obvious that we only need to prove one of the Equations (2) and (3), since on Ω

$$\limsup_{n\to\infty} \frac{S_n}{n} = -\liminf_{n\to\infty} \frac{-S_n}{n}.$$

In the following we will prove the Equation (2). It is trivial for $\overline{\mu} = \underline{\mu}$ and this theorem obviously holds true in this case from (I) of Theorem 1.1 of Chen and Wu [7] or Lemma 10. Hence we only need to consider $\overline{\mu} > \underline{\mu}$ By Lemma 10 or (I) of Theorem 1.1 of Chen and Wu [7], we have

$$V\left(\limsup_{n\to\infty}\frac{S_n}{n}>\bar{\mu}\right)=0.$$

Hence we only need to prove

$$V\left(\limsup_{n\to\infty}\frac{S_n}{n}\geq\bar{\mu}\right)=1.$$

For any subsequence $\{n_k\}_{k=1}^{\infty}$ of $\mathbb{N}$, we denote

$$A(a)=\left\{\omega\in\Omega:\limsup_{k\to\infty}\frac{S_{n_k}}{n_k}\geq a\right\},$$

$$A_{n_k}(a)=\left\{\omega\in\Omega:\frac{S_{n_k}}{n_k}\geq a\right\},\forall k\geq 1,a\in\mathbb{R}.$$

Since $\{X_n\}_{n=1}^{\infty}$ is a sequence of continuous random variables, thus $A(a)$ and $A_{n_k}(a)$ are both closed sets in F for all $a\in\mathbb{R}$ and $k\geq 1$. Thus $A_{n_k}^c(a)$ is an open set in F for any $k\geq 1$ and $a\in\mathbb{R}$. Then by the upper-continuity of V (see Proposition 1 (6)) for closed sets in F, we only need to prove for any fixed constant $\varepsilon\in(0,\bar{\mu}-\underline{\mu})$,

$$V\left(\limsup_{n\to\infty}\frac{S_n}{n}\geq\bar{\mu}-\varepsilon\right)=1.$$

Equivalently, we only need to prove for any fixed

$\varepsilon\in(0,\bar{\mu}-\underline{\mu})$,

$$v\left(\limsup_{n\to\infty}\frac{S_n}{n}<\bar{\mu}-\varepsilon\right)=0.$$

Then it is sufficient to find an increasing subsequence $\{n_k\}_{k=1}^{\infty}$ of $\mathbb{N}$ such that for any fixed $\varepsilon\in(0,\bar{\mu}-\underline{\mu})$,

$$v\left(\limsup_{k\to\infty}\frac{S_{n_k}}{n_k}<\bar{\mu}-\varepsilon\right)=0. \tag{4}$$

Noticing that

$$\begin{aligned}
&v\left(\limsup_{k\to\infty}\frac{S_{n_k}}{n_k}<\bar{\mu}-\varepsilon\right)\\
&\le v\left(\liminf_{k\to\infty}\frac{S_{n_k}}{n_k}<\bar{\mu}-\varepsilon\right)\\
&=v\left(\bigcap_{m=1}^{\infty}\bigcup_{k=m}^{\infty}A_{n_k}^{c}\left(\bar{\mu}-\varepsilon\right)\right)\\
&=\lim_{m\to\infty}v\left(\bigcup_{k=m}^{\infty}A_{n_k}^{c}\left(\bar{\mu}-\varepsilon\right)\right)\\
&=\lim_{m\to\infty}\left[1-V\left(\bigcap_{k=m}^{\infty}A_{n_k}\left(\bar{\mu}-\varepsilon\right)\right)\right]\\
&=1-\lim_{m\to\infty}V\left(\bigcap_{k=m}^{\infty}A_{n_k}\left(\bar{\mu}-\varepsilon\right)\right)\\
&=1-\lim_{m\to\infty}\lim_{l\to\infty}V\left(\bigcap_{k=m}^{l}A_{n_k}\left(\bar{\mu}-\varepsilon\right)\right),
\end{aligned} \tag{5}$$

since $A_{n_k}\left(\bar{\mu}-\varepsilon\right)$ are all closed sets and V is uppercontinuous for closed sets.

In addition, for any $m\ge 2, l\ge m$,

$$V\left(\bigcap_{k=m}^{l} A_{n_k}(\bar{\mu}-\varepsilon)\right)$$
$$\geq V\left(\bigcap_{k=m}^{l}\left[\{Y_k \geq \bar{\mu}-\varepsilon\}\cap\left\{\frac{S_{n_{k-1}}}{n_{k-1}} \geq \bar{\mu}-\varepsilon\right\}\right]\right)$$
$$= V\left(\bigcap_{k=m}^{l}\{Y_k \geq \bar{\mu}-\varepsilon\}\cap\left\{\frac{S_{n_{m-1}}}{n_{m-1}} \geq \bar{\mu}-\varepsilon\right\}\right)$$
$$\geq \mathbb{E}\left[\prod_{k=m}^{l} g^{\delta}(Y_k)\, g^{\delta}\left(\frac{S_{n_{m-1}}}{n_{m-1}}\right)\right],$$

where $Y_k = \dfrac{S_{n_k} - S_{n_{k-1}}}{n_k - n_{k-1}}$ and

$$g^{\delta}(x) = \begin{cases} 1, & x \geq \bar{\mu}-\varepsilon+\delta \\ \frac{1}{\delta}(x-\bar{\mu}+\varepsilon), & \bar{\mu}-\varepsilon < x < \bar{\mu}-\varepsilon+\delta; \\ 0, & x \leq \bar{\mu}-\varepsilon, \end{cases}$$

where δ is any fixed constant in $(0,\varepsilon)$. It is obvious that for any fixed $\delta \in (0,\varepsilon)$, g^{δ} is a bounded and Lipschitz continuous function on $\mathbb{R}$. Thus by the independence assumption we know that

$$V\left(\bigcap_{k=m}^{l} A_{n_k}(\bar{\mu}-\varepsilon)\right)$$
$$\geq \prod_{k=m}^{l} \mathbb{E}\left[g^{\delta}\left(\frac{S_{n_k} - S_{n_{k-1}}}{n_k - n_{k-1}}\right)\right] \mathbb{E}\left[g^{\delta}\left(\frac{S_{n_{m-1}}}{n_{m-1}}\right)\right]$$

And then by Lemma 9 for any fixed $m \geq 2$, and

$l \geq m$, if we choose a small constant $\tau = \dfrac{1}{(l+1-m)m}$, then there exists an integer $N = N(\delta, \varepsilon, l, m) \geq 1,$ such that for any $n_k - n_{k-1} \geq N$, we have

$$\mathbb{E}\left[g^{\delta}\left(\frac{\overline{S}_{n_k - n_{k-1}}}{n_k - n_{k-1}}\right)\right] \geq \sup_{\underline{\mu} \leq y \leq \overline{\mu}} g^{\delta}(y) - \tau = 1 - \tau,$$

$$m \leq k \leq l, \tag{6}$$

where we denote $\overline{S}_{n_k - n_{k-1}} := S_{n_k} - S_{n_{k-1}}$ for all fixed k with $m \leq k \leq l.$

Taking $n_k = N^k$ for any $m \leq k \leq l,$ we can obtain

$$V\left(\bigcap_{k=m}^{l} A_{n_k}(\overline{\mu} - \varepsilon)\right) \geq (1-\tau)^{l+1-m}\, \mathbb{E}\left[g^{\delta}\left(\frac{S_{n_{m-1}}}{n_{m-1}}\right)\right]. \tag{7}$$

Then letting l tend to ∞ and then letting m tend to ∞ on both sides of inequality (7), by Lemma 9 again, we can get

$$\begin{aligned}
&\lim_{m\to\infty} \lim_{l\to\infty} V\left(\bigcap_{k=m}^{l} A_{n_k}(\overline{\mu} - \varepsilon)\right) \\
&\geq \lim_{m\to\infty} \lim_{\tau\to 0^+} (1-\tau)^{\frac{1}{\tau m}}\, \mathbb{E}\left[g^{\delta}\left(\frac{S_{n_{m-1}}}{n_{m-1}}\right)\right] \\
&= \lim_{m\to\infty} \mathbb{E}\left[g^{\delta}\left(\frac{S_{n_{m-1}}}{n_{m-1}}\right)\right] \lim_{m\to\infty} \mathrm{e}^{-\frac{1}{m}} = 1
\end{aligned} \tag{8}$$

Thus from (5) and (8) we can obtain

$$v\left(\limsup_{k\to\infty} \frac{S_{n_k}}{n_k} < \overline{\mu} - \varepsilon\right) \leq 0.$$

Therefore, (4) holds true. We complete the whole proof. □

Remark 12. If $\underline{\mu}=\overline{\mu}=:\mu,$ then from the Theorem 11 we can see that $V\left(\lim_{n\to\infty}\frac{S_n}{n}=\mu\right)=1.$ This is just a trivial case for sequences without mean uncertainty.

Corollary 13. Let $\{X_n\}_{n=1}^{\infty}$ be a sequence of quasicontinuous random variables and satisfy all other conditions except for the continuity given in Theorem 11, then Theorem 11 still holds.

Proof. Similarly as the arguments in the proof of Theorem 11, we only need to prove

$$V\left(\limsup_{n\to\infty}\frac{S_n}{n}\geq\overline{\mu}\right)=1, \tag{9}$$

when $\overline{\mu}>\underline{\mu}.$

By the assumptions we know that for each $n\geq 1$ and any constant $2^{-n}>0,$ there exists an open subset O_n of Ω with $V(O_n)<2^{-n},$ such that X_n is continuous on O_n^c. Denote $O=\bigcap_{n=1}^{\infty}\bigcup_{k=n}^{\infty}O_k,$ then by Borel-Cantelli lemma (see Lemma 8) we can obtain

$$v(O)=V(O)=0.$$

For any $1>\delta>0,$ there exist an increasing subsequence $\{n_k\}_{k=1}^{\infty}$ of $\mathbb{N}$, an integer $N=N_\delta\geq 1$ and an open set $G^\delta\supset O$ (by Remark 7) satisfying

$$v\left(G^{\delta}\right) \leq v(O)+\delta=\delta,$$

such that when $n_1 \geq N,$ we have

$$G^{\delta} \supset \bigcup_{k=1}^{\infty} O_{n_k}.$$

Then X_{n_k} is continuous on $\left(G^{\delta}\right)^c$. By Lemma 6, for any $\varepsilon > 0$, we can find a compact set $K^{\varepsilon} \subset \left(G^{\delta}\right)^c$

with $a_{\varepsilon} = V\left(K^{\varepsilon}\right) \neq 0$ such that

$$a_{\varepsilon} = V\left(K^{\varepsilon}\right) \geq V\left(\left(G^{\delta}\right)^c\right) - \varepsilon \geq 1-\delta-\varepsilon.$$

Then we have

$$\begin{aligned}
& V\left(\limsup_{n\to\infty} \frac{S_n}{n} \geq \bar{\mu}\right) \\
& \geq V\left(\limsup_{k\to\infty} \frac{S_{n_k}}{n_k} I_{\left(G^{\delta}\right)^c} \geq \bar{\mu}\right) \\
& \geq \left(V \limsup_{k\to\infty} \frac{S_{n_k}}{n_k} I_{K^{\varepsilon}} \geq \bar{\mu}\right), \forall \varepsilon > 0.
\end{aligned} \tag{10}$$

For $\varepsilon > 0,$ we define

$$c_V^{\varepsilon}(A) = \frac{1}{a_{\varepsilon}} V(A), \forall A \in K^{\varepsilon} \cap F =: F^{\varepsilon}. \tag{11}$$

Then it is obvious that c_V^{ε} is a capacity on K^{ε} and satisfies all the properties of V given in Proposition 1 where Ω is substituted by K^{ε}. We also denote by H^{ε} the set of all random variables $X \in F^{\varepsilon}$ such that

$\mathbb{E}[X] < \infty$. Thus on $(K^{\varepsilon}, H^{\varepsilon}, \mathbb{E})$, $\{X_{n_k} I_{K^{\varepsilon}}\}_{k=1}^{\infty}$ is an independent and continuous sequence. Since K^{ε} is also a complete and separable metric space, by Theorem 11 we have

$$c_V^{\varepsilon}\left(\limsup_{k\to\infty} \frac{S_{n_k}}{n_k} I_{K^{\varepsilon}} \geq \overline{\mu}\right) = 1. \tag{12}$$

Then from (10)-(12) we have

$$V\left(\limsup_{n\to\infty} \frac{S_n}{n} \geq \overline{\mu}\right) \geq a_{\varepsilon} \geq 1 - \delta - \varepsilon. \tag{13}$$

Letting ε and δ tend to 0 in inequality (13) we can derive

$$V\left(\limsup_{n\to\infty} \frac{S_n}{n} \geq \overline{\mu}\right) \geq 1,$$

which implies (9). We complete the whole proof of this corollary.

4. EXTENSIONS

In Section 3 we get that the submean $\underline{\mu}$ and the supermean $\overline{\mu}$ are the inferior and superior limits of the arithmetic average of the first n random variables $X_1, X_2, \cdots, X_n$ given in Theorem 11, respectively, with probability 1 under the upper probability V. In fact, except the two values, any other value $x \in (\underline{\mu}, \overline{\mu})$ is still the limit of some subsequence of $\{X_n\}_{n=1}^{\infty}$, with probability 1 under V. We can see it in the following theorem.

Theorem 14. Under assumptions of Theorem 11, we have for any $x \in [\underline{\mu}, \overline{\mu}]$

$$V\left(x \in C\left(\frac{S_n}{n}\right)\right) = 1,$$

where $C(y_n)$ is a cluster of limit points of a real sequence $\{y_n\}_{n=1}^{\infty}$.

Proof. For $x = \underline{\mu}$ and $x = \overline{\mu}$, the result has been obtained in Theorem 11. For the trivial case, $\underline{\mu} = \overline{\mu}$, it is obvious. Now we consider $\overline{\mu} > \underline{\mu}$ and any $x \in (\underline{\mu}, \overline{\mu})$. We can notice that

$$V\left(x \in C\left(\frac{S_n}{n}\right)\right) = V\left(\liminf_{n\to\infty}\left|\frac{S_n}{n} - x\right| = 0\right)$$
$$= \lim_{\varepsilon\to 0} V\left(\liminf_{n\to\infty}\left|\frac{S_n}{n} - x\right| \leq \varepsilon\right),$$

where ε is any constant in $(0, \min\{\overline{\mu} - x, x - \underline{\mu}\})$,

since V is upper-continuous for closed sets. Thus we only need to find an increasing subsequence $\{n_k\}_{n=1}^{\infty}$ of $\mathbb{N}$ such that for any $\varepsilon \in (0, \min\{\overline{\mu} - x, x - \underline{\mu}\})$, we have

$$v\left(\limsup_{n\to\infty}\left|\frac{S_n}{n} - x\right| > \varepsilon\right) = 0, \tag{14}$$

Following the arguments in the proof of Theorem 11 we can obtain

$$v\left(\limsup_{n\to\infty}\left|\frac{S_n}{n}-x\right|>\varepsilon\right)$$
$$\le 1-\lim_{m\to\infty}\lim_{l\to\infty}V\left(\bigcap_{k=m}^{l}\{|Y_{k,x}|\le\varepsilon\}\cap\{|Z_{k,x}|\le\varepsilon\}\right)$$
$$\le 1-\lim_{m\to\infty}\lim_{l\to\infty}\prod_{k=m}^{l}\mathbb{E}\left[f^{\delta}(Y_{k,x})\right]\mathbb{E}\left[f^{\delta}(Z_{k,x})\right],$$

where $Y_{k,x}=\frac{S_{n_k}-S_{n_{k-1}}}{n_k-n_{k-1}}-x$ and $Z_{k,x}=\frac{S_{n_{m-1}}}{n_{m-1}}-x$ for any $k\ge 1$ and $x\in(\underline{\mu},\overline{\mu})$, δ is any given constant in $(0,\varepsilon)$ and

$$f^{\delta}(y)=\begin{cases}0, & y\ge\varepsilon;\\ \frac{1}{\delta}(\varepsilon-y), & \varepsilon-\delta<y<\varepsilon;\\ 1, & -\varepsilon+\delta\le y\le\varepsilon-\delta;\\ \frac{1}{\delta}(y+\varepsilon), & -\varepsilon<y<-\varepsilon+\delta;\\ 0, & y\le-\varepsilon.\end{cases}$$

Then by using the same arguments as in the proof of Theorem 11 we also can prove that (14) holds true. The whole proof is complete. □

The following corollary is obvious.

Corollary 15. Under the conditions of Theorem 11, for any continuous real function φ on $\mathbb{R}$, we have for all $x\in[\underline{\mu},\overline{\mu}]$,

$$V\left(\varphi(x)\in C\left(\varphi\left(\frac{S_n}{n}\right)\right)\right)=1.$$

In particular,

$$V\left(\limsup_{n\to\infty}\varphi\left(\frac{S_n}{n}\right)=\sup_{y\in[\underline{\mu},\overline{\mu}]}\varphi(y)\right)=1,$$

$$V\left(\liminf_{n\to\infty}\varphi\left(\frac{S_n}{n}\right)=\inf_{y\in[\underline{\mu},\overline{\mu}]}\varphi(y)\right)=1,\forall\varphi\in C(\mathbb{R}).$$

We also can extend Theorem 11, Theorem 14 and Corollary 15 to the sequences with different submeans and supermeans as follows.

Theorem 16. Let $\{X_n\}_{n=1}^{\infty}$ be an independent and continuous sequence under $\mathbb{E}$ and satisfy conditions

(1)-(3) of Lemma 9. Set $S_n=\sum_{i=1}^{n}X_i,\forall n\geq 1.$ Then for any $\varphi\in C(\mathbb{R})$ we have

$$V\left(\varphi(x)\in C\left(\varphi\left(\frac{S_n}{n}\right)\right)\right)=1,\forall x\in[\underline{\mu},\overline{\mu}].$$

Proof. By Lemma 10 and the proofs of Theorem 11 and Theorem 14 we only need to check whether (6) and (8) hold true under our assumptions of this theorem. In fact, from Lemma 9 they are obviously satisfied. Hence this theorem holds. □

From the proof of Corollary 13 and Theorem 16 we can immediately obtain the following corollary.

Corollary 17. Theorem 16 still holds when continuity assumption is substituted by quasi-continuity condition and condition (2) of Lemma 9 is replaced by the following condition:

(2') there exist real constants $\underline{\mu}\leq\overline{\mu}$ such that

$$\lim_{n\to\infty}\overline{\mu}_n=\overline{\mu},\lim_{n\to\infty}\underline{\mu}_n=\underline{\mu}.$$

5. AN EXAMPLE

Let $\Omega = C[0,T]$ with the supremum norm. Then Ω is a Banach space and compact, thus it is a separable and complete metric space with the distance generated by the norm of the space. Then we can define a G-expectation E_G, a special sublinear expectation (see Peng [4] for details), where $G(a) = \frac{1}{2}\left(a^+ - \sigma^2 a^-\right)$, σ is a nonnegative real number less than 1. Then for any bounded and independent sequence $\{X_n\}_{n=1}^{\infty}$ with the same submean $\underline{\mu}$ and supermean $\overline{\mu}$ in $L_G^1(F_T)$ under E_G, by Theorem 11 we have

$$V\left(\limsup_{n\to\infty} \frac{1}{n}\sum_{i=1}^{n} X_i = \overline{\mu}\right) = V\left(\liminf_{n\to\infty} \frac{1}{n}\sum_{i=1}^{n} X_i = \underline{\mu}\right) = 1,$$

where V is generated by E_G, since this sequence is a sequence of quasi-continuous random variables and from Denis, Hu and Peng [11] E_G can be represented as supremum of a family of linear expectations corresponding to a family of probabilities which is weakly compact.

6. ACKNOWLEDGEMENTS

This research is supported by WCU (World Class University) program through the National Research Foundation of Korea funded by the Ministry of Education, Science and Technology (R31-20007). The author gratefully thanks the referees for their careful reading.

REFERENCES

1. M. Marinacci, "Limit Laws of Non-Additive Probabilities and Their Frequentist Interpretation," Journal of Economic Theory, Vol. 84, No. 2, 1999, pp. 145-195.

2. F. Maccheroni and M. Marinacci, "A Strong Law of Large Numbers for Capacities," Annals of Probability, Vol. 33, No. 3, 2005, pp. 1171-1178.

3. Z. Chen, "Strong Laws of Large Numbers for Capacities," Unpublished, 2010. http://arxiv.org/abs/1006.0749

4. S. Peng, "Nonlinear Expectations and Stochastic Calculus under Uncertainty-with Robust Central Limit Theorem and G-Brownian Motion," Unpublished, 2012. http://arxiv.org/abs/1002.4546

5. Z. Chen, P. Wu and B. Li, "A Strong Law of Large Numbers for Non-additive Probabilities," International Journal of Approximate Reasoning, 2012. http://www.sciencedirect.com/science/article/pii/S0888613X12000783

6. F. Hu, "General Laws of Large Numbers under Sublinear Expectations," 2012. http://arxiv.org/abs/1104.5296

7. Z. Chen and P. Wu, "Strong Laws of Large Numbers for Bernoulli Experiments under Ambiguity," Nonlinear Mathematics for Uncertainty and Its Applications, Advances in Intelligent and Soft Computing, Vol. 100, 2011, pp. 19- 30. http://link.springer.com/chapter/10.1007%2F978-3-642-22833-9_2?LI=true

8. J. Xu and B. Zhang, "Martingale Property and Capacity under G-Framework," Electronic Journal of Probability, Vol. 15, No. 67, 2010, pp. 2041-2068.

9. G. Choquet, "Theory of Capacities," Annales de Institute Fourier, Vol. 5, 1954, pp. 131-295.

10. P. J. Huber and V. Strassen, "Minimax Tests and the Neyman-Pearson Lemma for Capacities," The Annals of Statistics, Vol. 1. No. 2, 1973, pp. 251-263.

11. L. Denis, M. Hu and S. Peng, "Function Spaces and Capacity Related to a Sublinear Expectation: Application to G-Brownian Motion Paths," Potential Analysis, Vol. 34, No. 2, 2011, pp. 139-161.

CHAPTER 6

Probability theory predicts That chunking into groups of three or four items increases the short-term memory capacity

Motohisa Osaka

Department of Basic Science, Nippon Veterinary and Life Science University, Tokyo, Japan

ABSTRACT

Short-term memory allows individuals to recall stimuli, such as numbers or words, for several seconds to several minutes without rehearsal. Although the capacity of short-term memory is considered to be 7 ± 2 items, this can be increased through a process called chunking. For example, in Japan, 11-digit cellular phone numbers and 10-digit toll free numbers are chunked into three groups of three or four digits: 090-XXXX-XXXX and 0120-XXX-XXX, respectively. We use probability theory to predict that the most effective chunking involves groups of three or four items, such as in phone numbers. However, a 16-digit credit card number exceeds the capacity of short-term memory, even when chunked into groups of four digits, such as XXXX-XXXX-XXXX-XXXX. Based on these data, 16-digit credit card numbers should be sufficient for security purposes.

KEYWORDS

Short-term Memory, Chunking, Probabilistic Model, Credit Card Number

1. INTRODUCTION

Short-term memory allows stimuli, such as numbers or words, to be recalled for several seconds to several minutes without rehearsal. Miller (1956) reported that the storage capacity of short-term memory was 7 ± 2 items, naming this "the magical number" [1] . He concluded that human "channel capacity" does not exceed a few bits and that unambiguous judgment of one-dimensional stimuli (i.e., all numbers) can be made from 7 ± 2 categories. Recently, Cowan (2001) reported that the capacity of short-term memory is 4 - 5 items [2] . Baddeley (1994) thought highly of "magical number seven", saying that it gives a beautifully clear account of information theory [3] , and several mathematical models investigating the origin of the magical number seven have been reported [4] [5] . Whether the capacity of short-term memory is 4 - 5 or 7 ± 2 items, it is clearly limited. However, memory capacity can be increased through a process called chunking [1] . For example, in Japan, 11-digit cellular phone numbers and 10-digit toll free numbers are chunked into groups of three or four digits: 090-XXXX- XXXX and 0120-XXX-XXX, respectively. Phone numbers in many other countries are similarly chunked. It is unclear how many items per group provide the most efficient chunking, and the current study used probability theory to investigate this.

In probability theory, there is a problem entitled "the tourist with a short memory" [6] . For example, if a tourist wants to visit four capitals A, B, C, and D, he travels first to one capital chosen at random. If he visits A, the next time, he should choose among B, C, and D with the same probability. However, in this problem, the tourist quickly forgets that he has already visited A. Therefore, if he visits B second, the next time, he would choose among A, C, and D with the same probability. The problem is to find the expected number, E(N), of trips required until the tourist has visited all four capitals. To address this question, the problem is transformed into a problem of short-term memory based on some hypotheses and assumptions described below. In the present paper these capitals correspond to items which are recalled. We study a case without chunking (Procedure 1), a case in which items are chunked in order into groups containing the same number of items (Procedure 2), and a case in which items are chunked in order into groups containing the different number of items (Procedure 3). The novelty of this study is that the most effective chunking involves groups of three or four items, such as in phone numbers, and that 23 trips may be the critical number, beyond which some items will be forgotten. A 16-digit credit card number exceeds the capacity of short-term memory, even when chunked into groups of four digits, such as XXXX-XXXX-XXXX-XXXX. Based on these data, 16-digit credit card numbers should be sufficient for security purposes.

2. MODEL

When a subject responds to an event involving several stimuli, those stimuli must be processed in such a way to distinguish among them while still associating them with the entire set of items. According to Miller's and Cowan's hypotheses (7 ± 2 or 4 - 5 items, respectively) [1] [2] , the capacity of short-term memory is between 4 - 9 items. Stimuli are often processed in order of dominance. The simplest way to order n items is to compare two items, retain the more dominant of the pair, then compare that with another item, again retaining the dominant one, and repeating this process until the entire collection has been ordered [5] . Although this process may be considered fundamental, it is assumed for simplicity that input items are one-dimensional categories, for example, strings of digits, letters, or words. The following assumptions were made:

Assumption 1: Input items (or stimuli) are assumed to be labeled as A_1, A_2, ..., and A_n in order.

Assumption 2: Items are remembered equally with no one item being more dominant. The probability to recall any A_j except A_i next after A_i is recalled is equal.

Assumption 3: The subject can only recall n items in order after he recalls every item at least once.

Applying these assumptions to the problem of "the tourist with a short memory", the problem is to find the expected number, E(N), of trips required until the tourist has visited all capitals. The process that any A_j except A_i is recalled after A_i is represented as a way: $W(A_i \rightarrow A_j)$. This can be calculated without chunking (Procedure 1) and with chunking into same-sized or different-sized groups (Procedures 2 and 3) as follows:

Procedure 1: To find the expected number, E(N), of ways required until all A_i's are recalled (Figure 1(a)).

Procedure 2: Items are chunked in order into groups, which have the same number of items (Figure 1(b)). For example, (A_1, A_2, A_3), (A_4, A_5, A_6), (A_7, A_8, A_9), ..., and (A_{n-2}, A_{n-1}, A_n). Groups are denoted in order as B_1, B_2, B_3, ... (Figure 1(b)). There is equal probability to recall any B_j except B_i immediately after B_i is recalled. When any B_j is recalled for the first time, all items in B_j are recalled at least once, which assumes that the relationship among the items in B_j has already been confirmed. Hence, all visits within B_j are remembered from the second visit of B_j onwards. When all B_i's are recalled, all A_i's are also recalled, confirming the relationship among all A_i's.

Procedure 3: Items are chunked into groups with different numbers of items. For example, in Japan, 11-digit cellular phone numbers and 10-digit toll free numbers are displayed as 090-XXXX-XXXX and 0120-XXX- XXX, respectively. The 11-digit phone number is chunked into three groups, B_1, B_2, B_3, one of which consists of three digits, B_1 = (A_1, A_2, A_3), and two of which consist of four digits, B_2 = (A_4, A_5, A_6, A_7), B_3 = (A_8, A_9, A_{10}, A_{11}). Similarly, the 10-digit toll free number is chunked into three groups, B_1, B_2, B_3, one of which consists of four digits, B_1 = (A_1, A_2, A_3, A_4), and two of which consist of three digits, B_2 = (A_5, A_6, A_7), B_3 = (A_8, A_9, A_{10}).

3. RESULTS OF CALCULATION

3.1. Procedure 1

When the number of all items is n, the expected number, E(N), of ways, $W(A_i \rightarrow A_j)$, required until all A_i's are recalled can be calculated.

In the case of n = 3, a subject wants to recall three items, A_1, A_2, A_3.

Set N as follows:

$$N = Y_0 + Y_1 + Y_2,$$

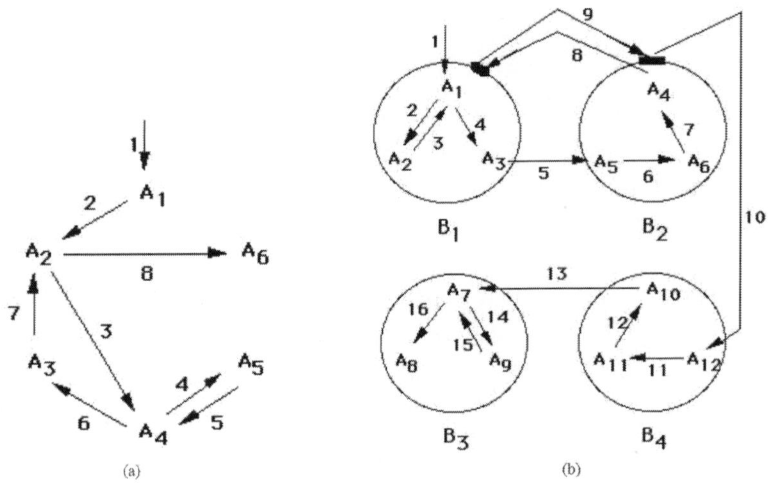

Figure 1. Ways, $W(A_i \rightarrow A_j)$, (labeled by turns) required until all A_i's are recalled without any chunking of items (a) and with chunking of items into, for example, four groups (B_1, B_2, B_3, B_4) (b).

where Y_m is the number of ways required for recalling one more item when m items have already been recalled. Therefore, Y_m's are independent stochastic variables. Y_0 and Y_1 are always 1. $Y_0 = 1$ indicates the first way of recalling one of the items. For example, it corresponds to the first way of Figure 1(a). In case of Y_2, one item has yet not been recalled, but it is recalled the kth time with a geometric probability of

$$p(1-p)^{k-1}; p = 1/2$$

for k = 1, 2, ... The expected distribution is 1/p. Therefore, $E(Y_2) = 2$. Since Y_m's are mutually independent random variables,

$$E(N)=E(Y_0)+E(Y_1)+E(Y_2)=1+1+2=4.$$

This equation is transformed into

$$E(N)=1+2\cdot\left(\frac{1}{2}+1\right).$$

When $Y_2 = y_2$, the probability of N; $P(N: Y_2 = y_2)$, is expressed as

$$P(N:Y_2=y_2)=P(Y_0=1)\cdot P(Y_1=1)\cdot P(Y_2=y_2).$$

$P(Y_0 =1)$ and $P(Y_1 = 1)$ are always 1.

$$P(Y_2=y_2)=\left(\frac{1}{2}\right)^{y_2-1}\cdot\frac{1}{2}=\frac{1}{2^{y_2}}.$$

Therefore,

$$P(N:Y_2=y_2)=\frac{1}{2^{y_2}}.$$

As $N=1+1+y_2$,

$$P(N:Y_2=y_2)=\frac{1}{2^{N-2}}.$$

Hence,

$$P(N)=\left(\frac{1}{2}\right)^{N-2}.$$

In the case of n = 4, a subject wants to recall four items, A_1, A_2, A_3, A_4.

Set N as follows:

$$N = Y_0 + Y_1 + Y_2 + Y_3,$$

where Y_i is the number of ways required for recalling one more item when i items have already been recalled. Therefore, Y_i's are mutually independent random variables. Y_0 and Y_1 are always 1. In the case of Y_2, two items have not yet been recalled, so one of these two items is recalled the kth time with a geometric probability of

$$p(1-p)^{k-1}; p = 2/3$$

for k = 1, 2, ... Similarly, Y_3 has a geometric probability function with p = 1/3. The expected distribution of a geometric probability function is 1/p. Therefore, $E(Y_2)$ = 3/2 and $E(Y_3)$ = 3. Since Y_i's are mutually independent random variables,

$$E(N) = E(Y_0) + E(Y_1) + E(Y_2) + E(Y_3) = 1 + 1 + \frac{3}{2} + 3 = \frac{13}{2}.$$

This equation is transformed into

$$E(N) = 1 + 3\left(\frac{1}{3} + \frac{1}{2} + 1\right).$$

Therefore, the expression for a general number, n, of items is:

$$E(N) = 1 + (n-1)\left(\frac{1}{n-1} + \frac{1}{n-2} + \cdots + \frac{1}{2} + 1\right).$$

This can be easily proven.

When $Y_2 = y_2$ and $Y_3 = y_3$, the probability of N; $P(N: Y_2 = y_2, Y_3 = y_3)$, is expressed as

$$P(N: Y_2 = y_2, Y_3 = y_3) = P(Y_0 = 1) \cdot P(Y_1 = 1) \cdot P(Y_2 = y_2) \cdot P(Y_3 = y_3).$$

$P(Y_0 = 1)$ and $P(Y_1 = 1)$ are always 1.

$$P(Y_2 = y_2) = \left(\frac{1}{3}\right)^{y_2 - 1} \cdot \frac{2}{3} = \frac{2}{3^{y_2}}.$$

$$P(Y_3 = y_3) = \left(\frac{2}{3}\right)^{y_3 - 1} \cdot \frac{1}{3} = \frac{2^{y_3 - 1}}{3^{y_3}}.$$

Therefore,

$$P(N : Y_2 = y_2, Y_3 = y_3) = \frac{2^{y_3}}{3^{y_2 + y_3}}.$$

As $N = 1 + 1 + y_2 + y_3,$

$$P(N : Y_2 = y_2, Y_3 = y_3) = \frac{2^{y_3}}{3^{N-2}}; 1 \le y_3 \le N - 3.$$

Hence,

$$P(N) = \sum_{y_3=1}^{N-3} \frac{2^{y_3}}{3^{N-2}} = \left(\frac{1}{3}\right)^{N-2} \left(2^{N-2} - 2\right).$$

In the case of n = 5, $N = 1 + 1 + y_2 + y_3 + y_4,$

$$P(N) = \sum_{y_3=1}^{N-4} \sum_{y_4=1}^{N-3-y_3} \frac{2^{y_3} \cdot 3^{y_4}}{4^{N-2}} = \left(\frac{1}{4}\right)^{N-2} \left(3^{N-2} - 3 \cdot 2^{N-2} + 3\right).$$

For a general number, n (≥3), of items,

$$P(N) = \left(\frac{1}{n-1}\right)^{N-2}$$

$$\left[\binom{n-2}{n-2} \cdot (n-2)^{N-2} + (-1)^1 \cdot \binom{n-2}{n-3} \cdot (n-3)^{N-2} + (-1)^2 \cdot \binom{n-2}{n-4} \cdot (n-4)^{N-2} + \cdots + (-1)^{n-3} \cdot \binom{n-2}{1} \cdot 1^{N-2}\right].$$

The equation is proved (Appendix). Specifically, in the case of n = 2, E(N) = 2 with a probability of 1; in the case of n = 3, E(N) = 4, and the cumulative probability that N is smaller than or equal to E(N), $P(N \le E(N))$, is 0.75; in the case of n = 4, E(N) = 13/2 and $P(N \le E(N)) = 0.7407$.

In the case of n = 5, which corresponds to one of Miller's magical numbers, $5(=7-2)$, E(N) = 28/3 ? 10 and in case of n = 9, which corresponds to the other of Miller's magical numbers, $9(=7+2)$, E(N) = 796/35 ? 23. In the case of n = 10, E(N) = 7409/280 ? 27. Clearly, as n increases, E(N) increases exponentially (Figure 2(a)). Hence, the greater the number, n, of items, the greater the difficulty to recall all items. Although the cumulative probability of $P(N \le E(N))$ decreases steadily, it is larger than 0.5 until n = 40 (Figure 2(b)).

3.2. Procedure 2

Items are chunked in order into groups with all groups containing the same number of items. The number of all items is denoted as n, and the number of items in each group is denoted as m. For an example of m = 3, the groups are (A_1, A_2, A_3), (A_4, A_5, A_6), (A_7, A_8, A_9), ... (A_{n-2}, A_{n-1}, A_n). These groups are denoted in order as B_1, B_2, B_3, ... (Figure 1(b)). Similar to Procedure 1, there is equal probability to recall any B_j except B_i immediately after B_i. When any B_j is recalled for the first time, all items in B_j are recalled at least once, so it is assumed that the relationship among the items in B_j has already been confirmed. Hence, all visits within B_j are saved from the second visit of B_j onwards. When all B_i's are recalled, it means that all A_i's are recalled, confirming the relationship among all A_i's.

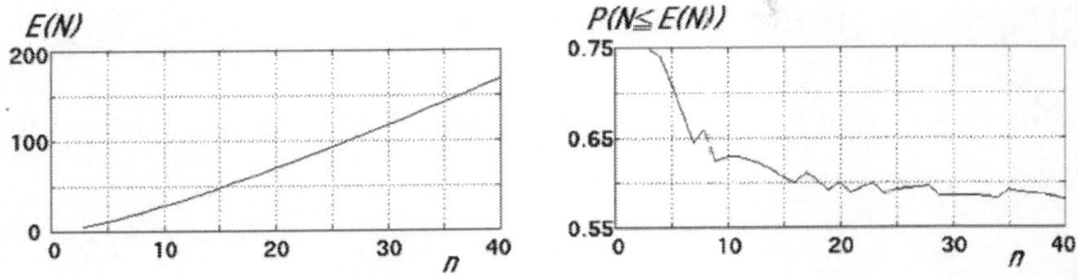

(a) (b)

Figure 2. (a) The expectation of the number, E(N), of ways, $W(A_i \rightarrow A_j)$, required until all A_i's are recalled; (b) The cumulative probability that N is smaller than or equal to E(N), P(N ≤ E(N)). n represents the number of items.

The number of B_i's is $\frac{n}{m}$, which is replaced by the nearest integer above $\frac{n}{m}$, $\left[\frac{n}{m}\right]$, if $\frac{n}{m}$ is not an integer.

The expected number, $E(N_{n,m})$, of ways required until all A_i's are recalled can be calculated. For the example of n = 12 and m = 3, a subject wants to recall 12 items, $A_1, A_2, A_3, \ldots, A_{12}$. Then, $B_1 = (A_1, A_2, A_3)$, $B_2 = (A_4, A_5, A_6)$, $B_3 = (A_7, A_8, A_9)$, and $B_4 = (A_{10}, A_{11}, A_{12})$.

Set $N_{12,3}$ as follows:

$$N = Z_0 + Z_1 + Z_2 + Z_3 + (Y_1 + Y_2) \times 4,$$

where Z_i is the number of ways required for recalling one more group when i groups have been recalled, and Y_j is the number of ways required for recalling one more item of any group when j items of this group have been recalled. Therefore, Z_i's and Y_j's are mutually independent random variables. Z_0, Z_1, and Y_1 are always 1. Specifically, $Z_0 = 1$ indicates the first way going to one of the groups. For example, it corresponds to the first way of Figure 1(b). Hence,

$$E(N_{12,3}) = E(Z_0) + E(Z_1) + E(Z_2) + E(Z_3) + [E(Y_1) + E(Y_2)] \times 4.$$

Using the case of four items in Procedure 1, we can regard the four groups in Procedure 2 as four items,

$$E(Z_0) + E(Z_1) + E(Z_2) + E(Z_3) = \frac{13}{2}.$$

Using the case of three items from Procedure 1,

$$E(Y_1) + E(Y_2) = 1 + 2 = 3.$$

Hence,

$$E(N_{12,3}) = \frac{37}{2}.$$

As another practical example, the expected number of ways required to recall 16 digits, $E(N_{16,4})$, corresponding to a credit card account number, XXXX-XXXX-XXXX-XXXX, can be calculated.

$$E(N_{16,4}) = E(Z_0) + E(Z_1) + E(Z_2) + E(Z_3) + [E(Y_1) + E(Y_2) + E(Y_3)] \times 4.$$

Using the case of four items in Procedure 1 and regarding the four groups as four items,

$$E(Z_0) + E(Z_1) + E(Z_2) + E(Z_3) = \frac{13}{2}.$$

Using the case of four items in Procedure 1,

$$E(Y_1) + E(Y_2) + E(Y_3) = \frac{11}{2}.$$

Hence,

$$E\left(N_{16,4}\right)=\frac{57}{2}.$$

Generally,

$$E\left(N_{n,m}\right)=\sum_{i=0}^{m}E\left(Z_i\right)+\left[\frac{n}{m}\right]\times\sum_{j=1}^{\left[\frac{n}{m}\right]}E\left(Z_i\right).$$

Then, if $\frac{n}{m}$ is an integer, $\left[\frac{n}{m}\right]=\frac{n}{m}$, otherwise $\left[\frac{n}{m}\right]$ stands for the nearest integer above $\frac{n}{m}$. $E(N_{n,m})$ can only be calculated precisely when n is a multiple of m. However, even if n is not a multiple of m, $E(N_{n,m})$ is calculated to observe the relationship between m and $E(N_{n,m})$. This calculation will be justified when n is larger than m, for example $n \geq 20$ and $1 \leq m \leq 10$. When $10 \leq n < 20$, $E(N_{n,m})$ is calculated only when n is a multiple of m. $E(N_{n,m})$ is calculated for n = 10, 11, ..., 100, and m = 1, 2, ..., 10. Figure 3 shows the results for n = 20, 30, 40, ..., 100 and m = 1, 2, ..., 10. When m = 3 and 4, $E(N_{n,m})$ is the smallest and the second smallest for any $n(10 \leq n \leq 100)$. When m = 2 or 5, $E(N_{n,m})$ is the third smallest. It is interesting to note that the case of m = 1 corresponds to any case without chunking from Procedure 1.

3.3. Procedure 3

The expected number $E(N_{n,*})$ of ways required until all A_i's are recalled can be calculated in the same manner as Procedure 2 for special cases of items chunked into groups of different lengths. When lengths of chunked groups, m = 2, 3, or 4, $E(N_{n,m})$ is the smallest. All integers are expressed by a sum of 2's, 3's, and 4's. For example, $17 = 2 + 3 + 4 \times 3$. Hence, items of any length can be chunked into groups, the lengths of which are 2, 3, or 4.

The 11-digit phone number 090-XXXX-XXXX is chunked into three groups, B_1, B_2, B_3, one of which consists of three digits, $B_1 = (A_1, A_2, A_3)$, and two of which consist of four digits, $B_2 = (A_4, A_5, A_6, A_7)$, $B_3 = (A_8, A_9, A_{10}, A_{11})$.

$$E(N) = E(Z_0) + E(Z_1) + E(Z_2) + \left[E(Y_1) + E(Y_2)\right] + \left[E(Y_1) + E(Y_2) + E(Y_3)\right] \times 2.$$

Hence,

$$E(N) = 4 + 3 + \frac{11}{2} \times 2 = 18.$$

The 10-digit phone number 0120-XXX-XXX is chunked into three groups, B_1, B_2, B_3, one of which consists of four digits, $B_1 = (A_1, A_2, A_3, A_4)$, and two of which consist of three digits, $B_2 = (A_5, A_6, A_7)$, $B_3 = (A_8, A_9, A_{10})$.

$$E(N) = E(Z_0) + E(Z_1) + E(Z_2) + \left[E(Y_1) + E(Y_2) + E(Y_3)\right] + \left[E(Y_1) + E(Y_2)\right] \times 2.$$

Hence,

$$E(N) = 4 + \frac{11}{2} + 3 \times 2 = \frac{31}{2}.$$

A 10-digit phone number of 03-XXXX-XXXX (for example, in Tokyo) is chunked into three groups, B_1, B_2, B_3, one of which consists of two digits, $B_1 = (A_1, A_2)$, and two of which consist of four digits, $B_2 = (A_3, A_4, A_5, A_6)$, $B_3 = (A_7, A_8, A_9, A_{10})$.

$$E(N)=E(Z_0)+E(Z_1)+E(Z_2)+E(Y_1)+\left[E(Y_1)+E(Y_2)+E(Y_3)\right]\times 2.$$

Hence, $$E(N)=4+1+\frac{11}{2}\times 2=16.$$

4. DISCUSSION

4.1. Findings Obtained from the Mathematical Model

4.1.1. Without chunking

As the number of the items, n, increases, the expected number, E(N), of ways required until all items are recalled increases exponentially. The cumulative probability that N is smaller than or equal to E(N), $P(N \leq E(N))$, decreases steadily. Hence, the greater the number, n, of items, the greater the difficulty to recall all items. In the case of five items, which corresponds to one of Miller's magical numbers (7 – 2 = 5), E(N) = 28/3 ? 10, and in the case of nine items, which corresponds to the other of Miller's magical numbers (7 + 2 = 9), E(N) = 796/35 ? 23. In the case of n = 10, E(N) = 7409/280 ? 27.

4.1.2. With chunking

$E(N_{n,m})$ is the expected number of ways required until all items are recalled. Hence, a smaller value for $E(N_{n,m})$ indicates more efficient recall. For example, the expected number of ways required until 12 items chunked into three groups are recalled, $E(N_{12,3})$, is 37/2 ? 19. In the case of a 16-digit credit card number, XXXX-XXXX- XXXX-XXXX, $E(N_{16,4})$ = 57/2 ? 29. From the results for n = 10, 11, ..., 100, and m =

1, 2, ..., 10, $E(N_{n,m})$ is the smallest for any n ($10 \le n \le 100$), when m = 3 or 4. Hence, when m = 3 or 4, all items can be recalled most quickly.

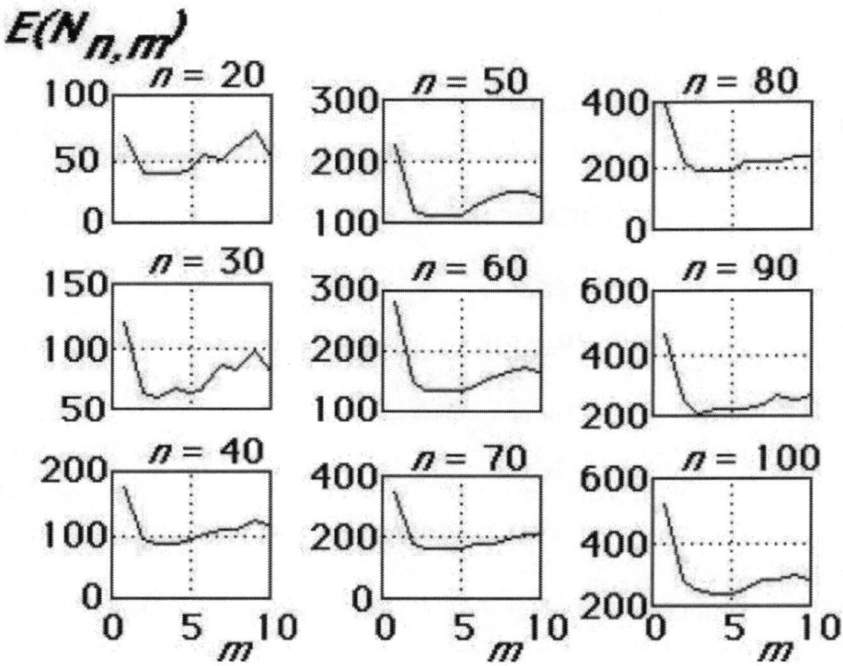

Figure 3. The expected number, $E(N_{n,m})$, of ways required until all A_i's are recalled. n represents the number of items. m represents the number of chunked groups.

4.1.3. Special cases of items chunked into groups of different lengths

The expected number of ways required to recall all 11 digits (e.g., in the phone number 090-XXXX-XXXX), $E(N_{11,*})$, is 18. For a 10-digit phone number in the format 0120-XXX-XXX, $E(N_{10,*}) = 31/2$? 16. For a 10-digit phone number in the format 03-XXXX-XXXX, $E(N_{10,*}) = 16$.

4.2. Interpretation of the findings

4.2.1. Without chunking

Short-term memory lasts from several seconds to several minutes. Based on the current data, we conclude that an individual can follow the 23 ways required to recall nine items within several minutes, but it takes longer to follow the 27 ways required to recall 10 items, so some one of the items are forgotten. These results suggest that 23 ways may be the critical number, beyond which some items will be forgotten.

4.2.2. With chunking

A smaller number of $E(N_{n,m})$ indicates more efficient recall. From the results for n = 10, 11, ..., 100, and m = 1, 2, ..., 10, $E(N_{n,m})$ is the smallest for any n, $(10 \leq n \leq 100)$ when m = 3 or 4. Each group has 3 or 4 items (m = 3 or 4) without chunking. From Procedure 1, $P(N \leq E(N))$ is 0.75 in the case of three items, and $P(N \leq E(N))$ is 0.7407 in the case of four items. $P(N \leq E(N))$ decreases steadily with more items. Hence, when m = 3 or 4, all items of each group can be recalled most quickly and with the greatest confidence. $E(N_{12,3}) = 37/2$? 19 is less than 23, the critical number for recall. Hence, chunking will be effective: $B_1 = (A_1, A_2, A_3)$, $B_2 = (A_4, A_5, A_6)$, $B_3 = (A_7, A_8, A_9)$, $B_4 = (A_{10}, A_{11}, A_{12})$. However, for 16 digits, such as in a credit card number, XXXX-XXXX- XXXX-XXXX, $E(N_{16,4}) = 57/2$? 29, which is larger than the critical number for recall. Thus chunking will not benefit short-term memory recall of a 16-digit credit card number. Based on these findings, a 16-digit credit card number of XXXX-XXXX-XXXX-XXXX should have greater security than a 12-digit number of XXX-XXX-XXX-XXX.

4.2.3. Special cases of items chunked into groups of different sizes

The expected numbers, E(N), of ways for 090-XXXX-XXXX, 0120-XXX-XXX, and 03-XXXX-XXXX, are less than 23, the critical number for recall. Hence, chunking into groups of two to four items is truly effective for recalling 11 or 10-digit phone numbers.

4.3. Study Limitations

The current findings were obtained using a model based on certain assumptions. The validity of these assumptions should be investigated in the future.

5. CONCLUSION

We use probability theory to predict that the most effective chunking involves groups of three or four items, such as in phone numbers, and conclude that an individual can follow the 23 ways required to recall nine items within several minutes, but it takes longer to follow the 27 ways required to recall 10 items, so some of the items are forgotten. These results suggest that 23 ways may be the critical number, beyond which some items will be forgotten. A 16-digit credit card number exceeds the capacity of short-term memory, even when chunked into groups of four digits, such as XXXX-XXXX-XXXX-

XXXX. Based on these data, 16-digit credit card numbers should be sufficient for security purposes.

REFERENCES

1. Miller, G.A. (1956) The Magical Number Seven plus or minus Two: Some Limits on Our Capacity for Processing Information. Psychological Review, 63, 81-97.

2. Cowan, N. (2001) The Magical Number 4 in Short-Term Memory: A Reconsideration of Mental Storage Capacity. Behavioral and Brain Sciences, 24, 87-114.

3. Baddely, A. (1994) The Magical Number Seven: Still Magic after All These Years. Psychological Review, 101, 353- 356.

4. Nicolis, J.S. and Tsuda, I. (1985) Chaotic Dynamics of Information Processing: The "Magic Number Seven Plus-Minus Two" Revisited. Bulletin Mathematical Biology, 47, 343-365.

5. Satty, T.L. and Ozdemir, M.S. (2003) Why the Magic Number Seven plus or minus Two. Mathematical and Computer Modelling, 38, 233-244.

6. Blom, G., Holst, L. and Sandell, D. (1991) Problems and Snapshots from the World of Probability. Springer-Verlag, New York.

CHAPTER 7

Some Applications of Bayes' Rule in Probability Theory to Electrocatalytic Reaction Engineering

Thomas Z. Fahidy

Chemical Engineering Department, University of Waterloo,
200 University Avenue West,Waterloo, ON, Canada

ABSTRACT

Bayesian methods stem from the principle of linking prior probability and conditional probability (likelihood) to posterior probability via Bayes' rule. The posterior probability is an updated (improved) version of the prior probability of an event, through the likelihood of finding empirical evidence if the underlying assumptions (hypothesis) are valid. In the absence of a frequency distribution for the prior probability, Bayesian methods have been found more satisfactory than distribution-based techniques. The paper illustrates the utility of Bayes' rule in the analysis of electrocatalytic reactor performance by means of four numerical examples involving a catalytic oxygen cathode, hydrogen evolution on a synthetic metal, the reliability of a device testing the quality of an electrocatalyst, and the range of Tafel slopes exhibited by an electrocatalyst.

1. INTRODUCTION

In a comprehensive overall treatment of the subject matter, Bockris and Khan [1] discuss electrocatalysis with respect to various physicochemical properties of substance and surface, for example, exchange current density, work function, bond strength, metal complexes, trace elements adatom effects, enzymatic catalysis, poisons, crystal face effects, and so forth, under the aegis of "phenomenological electrode kinetics". In the domain of ERE, the assessment of electrocatalyst (EC) performance also includes additional parameters related to catalyst preparation (i.e., possible defectiveness in specimens), cell construction, and human factors affecting reactor output.

This paper was written with this dichotomy in mind, from the vantage point of the electrochemical engineer, whose responsibilities dealing with production quota, the possibility of (temporary) reactor breakdown, safety, and environmental considerations reach well beyond purely scientific quantities. Major tools for dealing with these responsibilities are provided by probability-based (e.g., statistical) methods. Bayes' rule is one such tool, whose specific application to scenarios with EC is the subject of this article.

2. BRIEF THEORY

Following a concise definition [2] for the purposes of this paper, Bayes' rule for two events may be expressed as

$$P(A/B) = \frac{P(B/A)P(A)}{P(B)}, \tag{1}$$

where is the probability of event occurring if event has already occurred, and

$$P(B) = P(B/A)P(A) + P(B/A')P(A'), \tag{2}$$

which is the probability of event occurring, given the conditional probabilities (likelihoods) relating it to event and relating it to its opposite event . In an EC reactor a case for Bayes' rule would exist, for instance, when the loss of effectiveness in an EC may or may not be due to premature detachment of the catalytic layer. If is the event of detachment and is the event of deterioration (demise) of the EC, then would be the event of demise due to layer detachment, the event of a nondetachment cause of deterioration, and the event of deterioration due to a nondetachment cause. In terms of event probabilities, (1) yields the probability that deterioration would occur as a result of layer detachment and not due to a different cause, for example, the decomposition of a binder, or the splitting of the electrode frame, and so forth. If are mutually exclusive and exhaustive events, (1) and (2) are generalized to

$$P(A_k/B) = \frac{P(B/A_k)P(A_k)}{\sum_{k=1}^{n} P(B/A_k)P(A_k)}, \tag{3}$$

taking into account all possible causes of deterioration (the denominator of (3) is also known as the total probability theorem [3]). A lucid discussion of the merits of Bayesian methods by Bulmer [4] and a short set-theoretic proof by Arnold [5] are amply complemented by a sizeable literature on probability and statistics dealing with the subject matter.

Specific exploratory applications to electrochemical processes and technology at various levels of complexity are relatively recent [6–10]. The paper illustrates, via four independent examples, the (potential) utility of Bayes' rule in ERE. Due to the currently insufficient availability of appropriate experimental information in the research literature, hypothetical numerical data are employed with the sole purpose of indicating the course of analysis to which appropriate experimental data could be subjected. With the intention of stimulating at least a modest appetite at present for Bayes' rule, the illustrations are realistic but uncomplicated on purpose.

3. ILLUSTRATION OF THE UTILITY OF BAYES' RULE FOR ERE

3.1. Application No. 1: Estimating the Most Likely Location of Oxygen-Cathode Failure

Table 1 contains the failure-frequency map of identical oxygen cathodes, assumed to possess the structure described by Wiesener and Ohms [11]. These mutually independent failures are stipulated to have occurred in three independently operating electrochemical plants. Denoting as the source-of-failure events and as the plant location events, the probability of failure arising, for example, from human error is given by

$$P(A_5) = \sum_{k=1}^{3} P(A_5/B_k)P(B_k) = \frac{14}{48}\frac{48}{137} + \frac{11}{42}\frac{42}{137} + \frac{16}{47}\frac{47}{137}$$

$$= \frac{41}{137} = 0.2993, \tag{4}$$

which is about 30%, and Bayes' rule:

$$P\left(B_j/A_5\right) = \frac{P\left(A_5/B_j\right)P\left(B_j\right)}{P(A_5)}, \quad j = 1,2,3, \tag{5}$$

yields the probability of failure in any one of the three plants caused by human error: $P(B_1/A_5) = 14/41 = 0.3415;$; $P(B_2/A_5) = 11/41 = 0.2683; P(B_3/A_5) = 16/41 = 0.3802.$; . Thus, (next time) failure due to human error can be expected to be the least likely in Plant 2 and the most likely in Plant 3, although not significantly so with respect to Plant 1. The entire set $P(B_k/A_j),\ j = 1,2,3,\ k = 1,2,\ldots,5$, of likelihoods, obtained in a manner similar to (5) is shown in Table 2. The relatively largest failure probability, about 43%, can be expected in Plant 1 on account of insufficient PTFE binder content in the electrode layer. The contents of Table 2 would guide plant operators in attempting to eliminate (or at least to reduce the extent of) the most likely cause that can be expected in each plant. They would also indicate what cautionary measures would be advisable in the design of future plants.

Table 1. Postulated distribution pattern of 137 failure occurrences over a fixed time period in three independently operating (hypothetical) electrolytic plants using identical catalytic oxygen cathodes (Application 1).

Source of failure	Number of cathode failures over a fixed period of operation		
	Plant 1 (B_1)	Plant 2 (B_2)	Plant 3 (B_3)
A_1: Catalyst surface area[(1)]	12	14	10
A_2: Pore volume[(2)]	7	8	9
A_3: Binder content[(3)]	9	5	7
A_4: Catalyst content[(4)]	6	4	5
A_5: Human error[(5)]	14	11	16
Total number of failures	48	42	47

(1) Electrolytic carbon type P33: specific surface area is less than stipulated 1000 m^2/g.
(2) Electrolytic carbon type P33: specific pore volume is less than stipulated 2.3 cm^3/g.
(3) PTFE binder content in electrode layer is less than stipulated 10%.
(4) CoTAA (dibenzotetraazaannulen cobalt II) catalyst content on carbon is less than stipulated 15%.
(5) Careless stack assemblage and general operation.

Table 2. The complete set of probabilities computed via (5) in Application 1.

Source-of-failure events	$P(B_j/A_k)$		
	B_1	B_2	B_3
A_1	0.3333	0.3889	0.2778
A_2	0.2917	0.3333	0.3750
A_3	0.4286	0.2381	0.3333
A_4	0.4000	0.2667	0.3333
A_5	0.3415	0.2683	0.3902

3.2. Application No. 2: The Effect of Prior Probability on the Anticipated Viability of an EC-Generated H2 Evolution Process

A recently developed electrocatalyst for a hydrogen evolution process, made up of certain synthetic metals, is expected to possess an exchange current μA/cm2 at design operating conditions in a pilot scale electrolyzer. Inspection of Trasatti's [12] "volcano plot" [13, 14] suggests that its catalytic property would presumably fall between that of iridium and gold. It is further anticipated that the novel catalyst would be less expensive than Ir and Au, it would exhibit good dimensional/geometric stability as well as resistance to parasitic reactions due to possible contamination, and resistance to possible nonuniformity in current distribution. The design team postulates that if, on a pilot-plant scale, electrode specimens will show no loss in catalytic activity up to the passage of kAh/dm2 electric charge per unit area, then there should exist an a priori chance that a catalyst-carrying electrode (CCE), selected randomly from a lot of identically prepared specimens, can sustain its catalytic activity, at an acceptable level, up to the passage of kAh/dm2. During the tests, 91% of the electrodes were found to be acceptable, but 89% of electrodes, which later failed the tests, did not perform in a satisfactory manner. The design team (i) would proceed to consider commercial-scale implementation if there is at least a 95% chance that a survivor of the -test would keep its catalytic activity up to the passage of , (ii) would abandon further research if favourable results were obtained in only one-fifth or less of the tests.

The set of events of interest here, involving a randomly selected CCE, is defined as follows:

A: results obtained during the passage of are positive,

A': results obtained during the passage of are negative,

B: the CCE is acceptable,

B': the CCE is unacceptable,

A'/B: -results were positive for an acceptable CCE,

A/B'. Q1-results were positive for an unacceptable CCE,

A/B:: Q1-results were negative for an acceptable CCE,

A'/B'Q1 -results were negative for an unacceptable CCE,

B./A : Q1 a CCE which showed positive -test results is found acceptable,

B./A' : a CCE which showed negative -test results is found acceptable.

Consequently, the stipulations can be expressed in terms of their probabilities as follows: $P(A/B) = 0.91; P(A'/B') =$; $0.89; P(A/B') = 1 - - P(A'/B') = 0.11.$;. Bayes' rule yields, therefore,

$$P(B/A) = \frac{0.91P(B)}{0.91P(B) + 0.11[1 - P(B)]},$$

$$P(B/A') = \frac{0.09P(B)}{0.09P(B) + 0.89[1 - P(B)]}. \quad (6)$$

Here, is the prior probability of a CCE being acceptable. Its value, if not known experimentally, would be a matter of the designers' judgment. Table 3 indicates that in order to satisfy the $P(B/A) \geq$ ≥95% and $P(B/A')$ ≥20% decision criteria simultaneously, the prior probability of a CCE passing the -test would have to be somewhat higher than 70%. If the abandonment criterion were raised to a stricter $P(B/A')$ ≥25% probability, would have to be at least 77% for satisfying the two continuance conditions. Such results provide the design team with important knowledge for establishing proper testing protocols.

Table 3: The effect of prior probability on decision possibilities related to a new CE (Application 2).

$P(B)$%	$P(B/A)$%	$P(B/A')$%
20	67.4	2.5
40	84.6	6.3
60	92.5	13.2
70	95.1	19.1
80	97.1	28.8
90	98.7	47.6

$P(B/A)$: the probability that a CE will be deemed acceptable upon the Q_2 test, if the results of the Q_1 test were positive.
$P(B/A')$: the probability that a CE will be deemed acceptable upon the Q_1 test, even if the results of the Q_1 test were negative.

3.3. Application 3: Probing Claims Regarding the Reliability of a Catalyst Tester

A device for testing defects in a certain electrocatalyst (EC) is envisaged to be advertised by the catalyst producer, claiming that it is 97% reliable if the EC is defective, and 99% reliable when it is flawless. Independently from any testing device and based upon earlier experience, 4% of said EC may be expected to be defective upon delivery. In order to ascertain the true reliability of the device, Bayes' rule is applied to basic event set : the EC is defective; : the EC is flawless; : the EC is tested to be defective; : the EC is tested to be flawless, equipped with the full set of conditional events of interest here with their probabilities::

B/A: EC is (known to be) defective, and tested defective, ,

B' /A: : EC is (known to be) defective, but tested flawless, ,:

B/A' EC is (known to be) defective, but tested defective, ,

B' /A:: EC is (known to be) flawless, and tested flawless, .

The probabilities of events to be computed via Bayes' rule, shown in Table 4, point to the (vexingly) high possibility of rejecting flawless EC's (about 20%) and the (vexingly) low possibility of identifying defective EC's (about 80%) when the tester indicates defectiveness. These findings, hidden by the advertisement without Bayes' rule, should discourage its adoption for routine use.

Table 4: Probabilities of flawlessness/defectiveness expected from an EC tester (Application No. 3).

Event	Bayes' rule	Event probability
EC tested defective, but found flawless	$P(A'/B) = \frac{P(B/A')P(A')}{P(B/A')P(A') + P(B/A)P(A)}$	0.1983
EC tested flawless, and found flawless	$P(A'/B') = \frac{P(B'/A')P(A')}{P(B'/A')P(A') + P(B'/A)P(A)}$	0.9987
EC tested flawless, but found defective	$P(A/B') = \frac{P(B'/A)P(A)}{P(B'/A)P(A) + P(B'/A')P(A')}$	0.0013
EC tested defective, and found defective	$P(A/B) = \frac{P(B/A)P(A)}{P(B/A)P(A) + P(B/A')P(A')}$	0.8017

$P(A/B') + P(A'/B') = P(A/B) + P(A'/B) = 1$; $P(A) = 0.04$; $P(A') = 1 - 0.04 = 0.96$.

3.4. Application 4: Probing Claims Regarding Tafel Slopes in an Electrocatalytic Oxidation of Methanol Process Envisaged for Fuel Cells

This example is motivated by an experimental study of Pt:Mo dispersed catalysts (PMDCs) for the electro-oxidation of methanol in acid medium [15], assuming that a different research team claims in a new experimental catalyst development program a 65–70 mV/dec Tafel slope range at low current densities, and a 255–265 mV/dec at high current densities (in contrast with the 30–35 and 230–250 mV/dec ranges, resp. in the cited study). The polarization method chosen for investigating the claim is assumed to be 89% reliable when the claim cannot be verified and 99.5% reliable when the claim can be verified. Defining events : the PMDC exhibits Tafel slopes below the claimed ranges and : the PMDC is found to exhibit Tafel slopes below the claimed ranges, the complementary events : the PMDC exhibits Tafel slopes within the claimed ranges and : the PMDC is found to exhibit Tafel slopes within the claimed ranges establish the basis for applying Bayes' theorem. Following the pattern shown by the previous applications, the conditional probabilities are , $P(B'/A) = 0.11, P(B/A') = 0.005, P(B/A) = 0.89$, and . $P(B'/A') = 0.995.$

The research team is assumed to report that 96% of the new PMDC possess the claimed Tafel slope ranges; Bayes' theorem yields $P(A'/B) = 0.1188$; $P(A/B') = 0.0046$; $P(A'/B') = 0.9954$; $P(A/B) = 0.8812.$; ; ; . The about 12% probability that a new catalyst complies with the claim although the polarization experiment indicates otherwise raises at least a reasonable doubt about the claim or the reliability of the experimental procedure, in spite of the satisfactory $P(A/B')$ and $P(A'/B')$ values.

4. DISCUSSION AND FINAL REMARKS

Perhaps the most striking feature of Bayes' rule is the amount of information that can be gleaned from a few uncomplicated probability ratios (the fact that Bayesian methods are at present more than two hundred years old is equally impressive). Within the Bayesian framework, a prior event probability is updated to a posterior probability of that event by means of a likelihood. The latter provides the (conditional) probability of corroborating the a-priori stated hypothesis; this aspect is numerically illustrated in the Appendix.

The examples presented in this paper provide a small "window" to the realm of Bayesian methods whose further exploration in electrochemical science and engineering requires further work. Bayes' rule is just one of many other mathematical devices of applied probability theory with potential interest to the field.

ACKNOWLEDGMENTS

Utilities provided by the University of Waterloo and the Natural Sciences and Engineering Research Council of Canada (NSERC) are gratefully acknowledged.

REFERENCES

1. J. O'M. Bockris and S. U. M. Khan, Surface Electrochemistry, Plenum, New York, NY, USA, 1993.

2. R. Porkess, Collins Internet-Linked Dictionary of Statistics, Harper Collins College, Glenview, Ill, USA, 2nd edition, 2005.

3. E. B. Manoukian, Modern Concepts and Theorems of Mathematical Statistics, Springer, New York, NY, USA, 1986.

4. M. G. Bulmer, Principles of Statistics, Dover, New York, NY, USA, 2nd edition, 1979.

5. S. F. Arnold, Mathematical Statistics, Prentice Hall, Englewood Cliffs, NJ, USA, 1990.

6. T. Z. Fahidy, "On the utility of certain decision theory techniques in electrochemical process design," in Current Topics in Electrochemistry, vol. 7, pp. 119–123, Research Trends, Trivandrum, India, 2000.

7. T. Z. Fahidy, "An application of Bayesian risk theory to electrochemical processes," Electrochimica Acta, vol. 49, no. 9-10, pp. 1397–1401, 2004.

8. T. Z. Fahidy, "Bayesian updating of electrochemical process parameters via natural conjugate probability distributions," Electrochimica Acta, vol. 49, no. 27, pp. 5013–5021, 2004.

9. T. Z. Fahidy, "Probabilstic methods in the analysis of certain electrochemical systems," in Recent Research Developments in Electrochemistry, S. G. Pandalai, Ed., vol. 6, pp. 116–117, Transworld Research Network, Trivandrum, India, 2003.

10. T. Z. Fahidy, "Electrochemical applications of net-benefit analysis via Bayesian probabilities," Journal of Applied Electrochemistry, vol. 37, no. 6, pp. 747–752, 2007.

11. K. Wiesener and D. Ohms, "Electrocatalysis for environmentally orientated electrochemical processes and environmental protection," in Environmental Oriented Electrochemistry, C. A. C. Sequeira, Ed., p. 694, Elsevier, Amsterdam, The Netherlands, 1994.

12. S. Trasatti, "Work function, electronegativity, and electrochemical behaviour of metals. III. Electrolytic hydrogen evolution in acid solutions," Journal of Electroanalytical Chemistry, vol. 39, no. 1, pp. 163–184, 1972.

13. J. O'M. Bockris and S. U. M. Khan, loc. cit., Section 3.14.3, Fig. 3.23, p. 268.

14. J. O'M. Bockris and S. U. M. Khan, loc. cit., Section 3.21.6, Fig. 3.31, p. 293.

15. Oliveira Neto, J. Perez, W. T. Napporn, E. A. Ticianelli, and E. R. Gonzalez, "Electrocatalytic oxidation of methanol: study with Pt:Mo dispersed catalysts," Journal of the Brazilian Chemical Society, vol. 11, no. 1, pp. 39–43, 2000.

CHAPTER 8

A new RREQ message forwarding technique based on Bayesian probability theory

Venetis Kanakaris*, David L Ndzi, Kyriakos Ovaliadis and Yanyan Yang

School of Engineering, University of Portsmouth, Anglesea Road, Portsmouth, PO1 3DJ, UK

ABSTRACT

The flooding method, which is used by many mobile ad-hoc routing protocols, is a process in which a route request packet (RREQ) is broadcasted from a source node to other nodes in the network. This often results in unnecessary re-transmissions, causing packet collisions and congestion in the network, a phenomenon called broadcast storm. This article presents firstly the impact of a different message forwarding probability on the RREQ and secondly a RREQ message forwarding scheme which is implemented on Ad-hoc On-

Demand Distance Vector Routing (AODV) routing protocol, a Bayesian probability based the AODV extended version based on a modified version of Bayesian probability (AODV_EXT_BP) that reduces routing overheads, by calculating the probability with respect to the neighbour density as well as the posterior probability. The performance of the AODV_EXT_BP is compared to that of extended version of AODV (AODV_EXT), AODV, Destination Sequenced Distance Vector, dynamic source routing and Optimized Link State Routing protocols and the simulation results show that the AODV_EXT_BP protocol achieves better results in all sectors.

KEYWORDS

Mobile ad-hoc network Routing message overhead Route discovery Broadcast Bayesian probability Flooding Power-aware routing Routing protocols AODV AODV_EXT AODV_EXT_BP DSDV DSR OLSR

1. INTRODUCTION

An ad-hoc network is a promising technology which can be applied in an extensive number of areas ranging from environmental monitoring to disaster management. Furthermore, ad-hoc networks can be implemented in application of sensors for process automation in a diversity of industrial applications. Events such as earthquakes can often serve to illustrate the weakness of centrally managed networks and the importance of research and development in ad-hoc networks such as mobile ad-hoc networks (MANETs), where a centralized connectivity is not needed. MANET is a wireless network which consists of mobile nodes with no pre-determined infrastructure. One of the major restrictions of ad-hoc network systems is the energy availability and continuous reduction in the size of devices which means that power reduction cannot simply be improved with big battery systems[1]. Apart from the operation of onboard electronics, power consumption depends on numerous processes and overheads required to maintain connectivity.

There are numerous recommended solutions to deal with connectivity problems and power limitation in ad-hoc networks. Such techniques comprise development of hardware, routing algorithms, protocols and battery technology or power management systems[2, 3]. There are researchers that have suggested the development of an optimized hardware which can be used in applications based on data rates[4]. There is a proposal in[5, 6] to adapt energy control to applications where the voltage, and therefore processing speed and power, can be reduced for non-time sensitive applications. Other methods that have been proposed in[7] are aimed at avoiding network partitioning by controlling power consumptions of critical link nodes. Recently, many probabilistic approaches have been proposed in[8–11] with the aim to alleviate the flooding phenomenon and solve the broadcast storm problem. The authors have already proposed a probability-based algorithm in[12] which is an improved version of Ad-hoc On-Demand Distance Vector Routing (AODV), the AODV_EXT, which reduces energy consumption by 3%. In this article, we propose a modification of AODV, the AODV_EXT_BP, protocol which improves the energy and data transmission efficiency of the network by using the Bayesian probability theory and give better results compared to AODV_EXT.

In general, ad-hoc wireless networks broadcast packets to the whole network as a means of transmitting information from one node to the other in the network[13]. Broadcasting in MANETs is not only an essential action for unicast routing protocols in mobile scenarios, but also a unique process of a number of multicast routing protocols. A variety of unicast, multicast, and geocast protocols uses the broadcasting method in order to provide the important control and route establishing functionality. Broadcasting a packet to the whole network has a wide range of applications in MANETs. Consequently, improving the process of broadcasting will result in savings in several ad-hoc applications.

Flooding is the simplest technique used by source nodes to broadcast packets to the adjacent nodes[14]. Each neighbour node receiving the packet for the first time rebroadcasts it ensuring forward propagation from the source node until every node in the network has received and transmitted the broadcast packet exactly once.

The broadcasting protocols can be classified into two main categories, namely deterministic and probabilistic. The probabilistic approach usually offers a simple solution in which every node that receives a broadcast packet has a predefined probability of forwarding the message[15]. But this approach does not ensure full network coverage. On the other hand, the deterministic approach can provide full coverage and can be further categorized into two categories, namely location information and neighbour set based.

In MANETs, the routing task is delivered through network nodes which act as both routers and end points in the network. In order for a route to a specific destination node to be discovered, existing on-demand routing protocols use a simple flooding mechanism whereby a route request packet (RREQ) originating from a source node is broadcasted without exception to all nodes in the network[16]. This can result to considerable redundant retransmissions, causing congestion and packet collisions in the network.

The implication of broadcast routing messages on network performance is presented in this article. The following protocols have been studied and their performances in simulated networks are analysed: dynamic source routing (DSR), AODV, Destination Sequenced Distance Vector (DSDV) routing and Optimized Link State

Routing (OLSR). These protocols have widely been used and cited in literature[17].

This article is organized as follows: "Routing protocols" section describes the protocols that will be evaluated in this article. The routing procedure of AODV is also described. "The modification of AODV using different RREQ mechanism message forwarding probability scheme" and "Simulation and metrics of a different message forwarding probability scheme on the RREQ mechanism of AODV protocol" sections present the modification and the results of the AODV behaviour when using different message forwarding probability implemented on the RREQ mechanism, respectively. "The modification of AODV RREQ mechanism using Bayesian probability approach" section presents a description of the proposed modification to AODV as well as the Bayesian probability theory. In "Simulation and metrics of AODV RREQ mechanism using the Bayesian approach" section, the simulated scenario is described together with the settings, network configurations and the parameters that have been used in order to evaluate the performance of the protocols. The results of the simulation and discussions are provided in "Results and discussions" section which is followed by the conclusions.

2. ROUTING PROTOCOLS

In DSDV[18] protocol, there is an exchange of messages between mobile nodes within range. Routing updates may be enabled or routine. The update process starts when routing information from one of the adjacent nodes forces a change in the routing table. If there is a packet which the route to its destination is unknown, it is cached whilst routing queries are sent out. The packets are cached until route-replies are received from the destination node. The buffer has a specific size and time limitation for caching packets beyond which packets are dropped. When the route to the destination is known the packets are routed directly. In a case that the destination node is not found, the packets are forwarded to the default target node which is the routing agent.

In DSR protocol, the routing agent controls every data packet for source-route information[19]. The packets are then forwarded according to the routing information. In case, it cannot find any routing information in the packet, it offers the source route if route is known. When the destination is not known it caches the packet and broadcasts route queries. The routing query is initially sent to all the adjacent nodes and it is always enabled by a data packet which has no route information about its destination. Route-replies are sent back if the information regarding the route to the destination is needed.

The OLSR[20] is a routing protocol in which the nodes are aware of all the valid routes. The OLSR protocol uses the flooding method in order to inform all active nodes in the network when there is a topology change. The OLSR reduces the possible overhead in the network by using the multi-point relays (MPR). The main concept of MPR[21] is to reduce the number of duplicate retransmissions when a broadcast packet is forwarded. By this method, the number of retransmissions is confined to a small group of nearby nodes, instead of using all the neighbours. This group of nodes is kept as small as possible by selecting the nodes which cover (in terms of one-hop radio range) the same network region as the complete set of nearby nodes. The OLSR routing protocol has two types of control messages, namely Topology Control (TC) and Hello. TC messages are used for sending information regarding the adjacent nodes which contains the MPR selector list, instead of Hello messages which are used in order to discover information about the link status and the host's neighbours. The OLSR protocol has a drawback due to the fact that every host periodically transmits the updated topology information to the whole network thus increasing bandwidth usage. But this problem is solved by using the MPR, which forwards only the messages regarding the topology of the network.

The AODV protocol is a mix of DSDV and DSR protocols[22]. It uses the hop-by-hop routing sequence numbers and beacons of DSDV and keeps the basic route-discovery and route-maintenance mechanism of DSR. When a node needs to know a route to a specific destination, it generates an RREQ. The RREQ packet is forwarded by intermediate nodes which also create a reverse route from the destination. When the request reaches a node with a route to the destination node it also creates a Route Reply (RREP) which contains the number of hops that are required to reach the destination. The intermediate nodes that take part in forwarding, reply to the source node and create a forward route to destination. This route created from each node from source to destination is a hop-by-hop state and not the entire route as in source routing.

3. THE MODIFICATION OF AODV USING DIFFERENT RREQ MECHANISM MESSAGE FORWARDING PROBABILITY SCHEME

In this article, we propose a modification to AODV protocol to use a probability-based message forwarding scheme in the RREQ mechanism. In the original AODV protocol, in the RREQ mechanism, 100% (probability of 1.0) all intermediate nodes are involved in forwarding the message in order to find a route to a destination. In our proposed modification, we can have the same outcome by using only a percentage of the number of the intermediate nodes. By using a lower probability than 1.0, we use only a selected number of nodes to forward the messages thus reducing energy consumption and the number of retransmissions. The following lines describe the proposed modification to the RREQ mechanism:

- Assume that there are N nodes in the network and n is the number of nodes in the neighbourhood of a transmitting node;
- In normal AODV route message forwarding, if a node is forwarding a packet then all the n neighbours will try to forward the message again. If P_{rx} is the power consumed by a node when receiving a route control message and P_{tx} is the power consumed when a node forwards a route control message, then all n nodes will spend

$$P_s = (P_{rx} + P_{tx}) * n \tag{1}$$

- If P_i is the sum of total initial power of the n neighbouring nodes, then the total remaining power of the neighbouring node remaining is

$$P_{\text{total}} = (P_i - P_s) \tag{2}$$

- If probabilistic route message forwarding scheme is used, for example, assume that only 50% of the nodes are allowed to forward the message at any instance based on probability of 0.5, this means that only$n$2
- nodes will receive and forward the message. This means that the network residual power will be

$$P_{\text{total}} = \left(P_i - \frac{P_s}{2}\right) \tag{3}$$

- If we compare Equations (2) and (3), the latter will preserve battery power and double the lifetime of the network whilst reducing the number of retransmissions.

4. SIMULATION AND METRICS OF A DIFFERENT MESSAGE FORWARDING PROBABILITY SCHEME ON THE RREQ MECHANISM OF AODV PROTOCOL

Network Simulator 2 (NS2) has been used to evaluate the modified AODV routing protocol using the probability-based message forwarding scheme (Table1). The simulations were carried out to assess the performance of the routing protocols with network sizes of 10, 20, 30, 40, 50, 60, 70, 90 and 100 nodes with mobile node speeds between 1 and 30 m/s. The mobility mode that has been used in all simulation was the random way point. In this model, nodes in a specific area choose some destination, and move there at a random speed uniformly chosen from (0, $V_{\max}$), where $V_{\max}$ is the maximum speed of the simulation. The average speed is expected as the simulation evolves, and simulation results are in the form of an average over a period of time[23]. The simulation scenario has been repeated ten times and hence the results are the average of the ten times simulation scenarios. In all cases, the nodes send Constant Bit-rate (CBR) over user datagram protocol (UDP). The metrics that have been used to evaluate the performance of the network and protocols are the following[24]:

- *Consumed power*: the average consumed battery power.
- *Number of packets dropped:* this is the number of data packets that are not successfully sent to its destination.
- *MAC Load:* this is the ratio of the number of MAC layer messages propagated by every node in the network to the number of data packets successfully delivered to all destination nodes. In other words, the MAC Load is the average number of MAC messages generated for each data packet successfully delivered to the destination.

Table 1 . Definition Of The Different Message Forwarding Probability

P_{tx}	Power consumed during forwarding a route control message
P_{rx}	Power consumed during receiving a route control message
P_{total}	The sum of total remaining power of the neighbouring nodes node
P_i	The sum of total initial powers of the n neighbours
P_s	The total power of all nodes
n	Number of nodes in the neighbourhood

The simulation configuration and specification are specified in Table2.

Figure1 shows that 0.3 (30% of the intermediate nodes) and 0.4 (40% of the intermediate nodes) forward probability in terms of power consumption present very good results in every network size instead of using probability 1.0 which is the original probability that the AODV protocol uses. Moreover, good results are achieved by 0.2 probability (20% of the intermediate nodes) but not in every network size. As we can see from the graph (Figure1) in large-scale networks (more than 70 nodes), the use of 0.2 forward probability consumes more power than the 1.0 forward probability, meaning that needs more retransmissions in order to a valid route to the destination node to be found. By using less probability than 1.0 (100%) we use only a selected number of nodes to forward the messages avoiding with this way the consumption of more energy.

Table 2. Simulation Parameters Of Aodv To Evaluate Different Rreq Message Forwarding Probabilities

The protocol	Probabilistic RREQ message forwarding protocol
Topographical area	800 × 800 m^2
Number of nodes	10, 20, 30, 40, 50, 60, 70, 80, 90, 100
Mobility	1–30 m/s
Probability	0.10–1.0 (step 0.1)
Radio-propagation model	TwoRayGround
Network interface type	WirelessPhy
MAC type	802_11
Interface queue type	DropTail/PriQueue
Antenna model	OmniAntenna
Total simulation time	100 s
Transport/traffic type	CBR over UDP
Tx power of the nodes	0.1819 W
Rx power of the nodes	0.0501 W
Idle power of the nodes	0.0350 W
Initial energy of the nodes	1000.0 J
Scenario simulation repetitions	10

Figure2 shows that in small network sizes, smaller than 30 nodes, the forward probability-based scheme presents almost identical results, whilst in larger network sizes (bigger than 30 nodes) the use of 0.3 or 0.8 forwarding probability shows lower dropped packets than using probability 0.1, 0.7 and 1.0. This shows that when using of the following forwarding probabilities, 0.3 and 0.8, the number of retransmissions is lower as the number of data packets that are successfully sent to their destinations are higher. Good results are also achieved with 0.5 forwarding probability but only for network sizes with the number of nodes between 70 and 100.

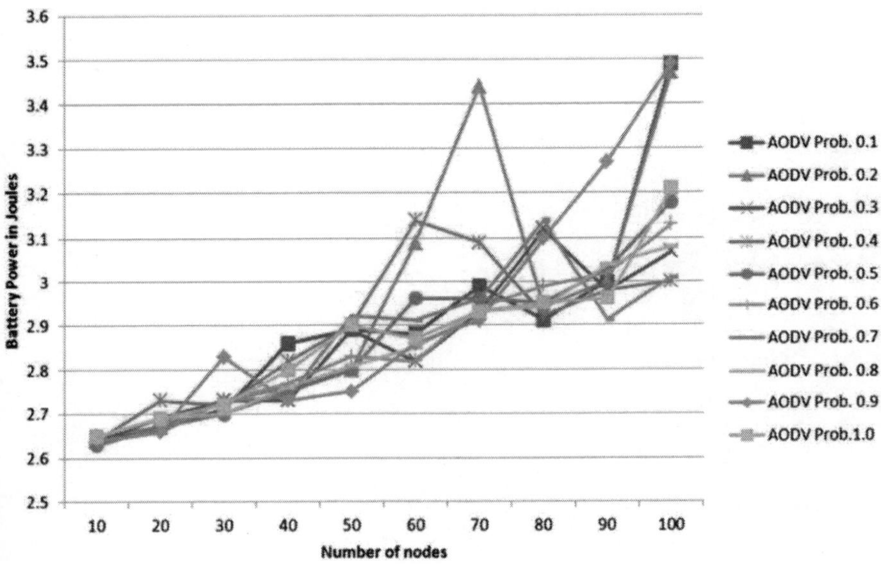

Figure 1. The Average Consumed Power.

Figure 3 shows that the MAC Load increases rapidly when the probability is 0.1, 0.7 and 1.0 while on probability 0.3, 0.4 and 0.8 the MAC Load is very low. The MAC Load explains how successfully the number of data packets is delivered to the destination node. The forwarding probabilities 0.3, 0.4 and 0.8 provide very low MAC Load meaning high success on data packet delivery while all the other forwarding probability values give very high MAC Load which lead to redundant retransmissions and depletion of battery power of the intermediate nodes.

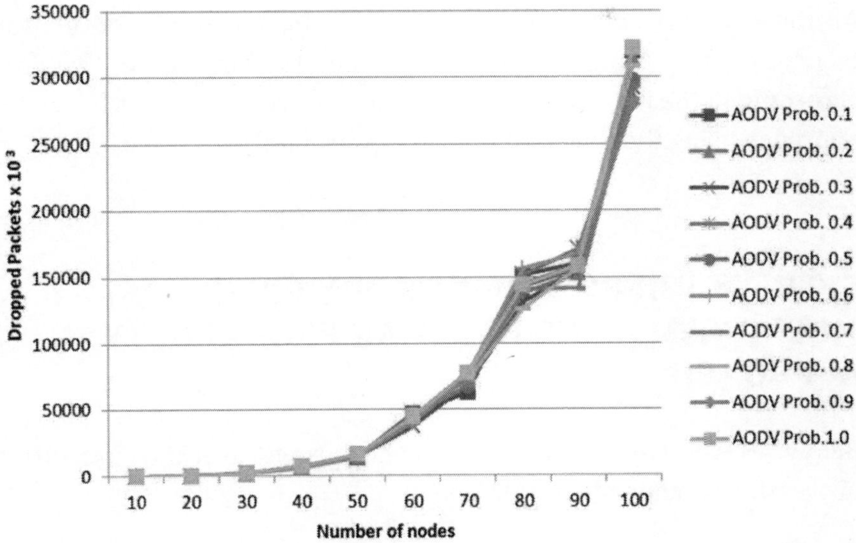

Figure 2 . Dropped Packets In Terms Of Network Size.

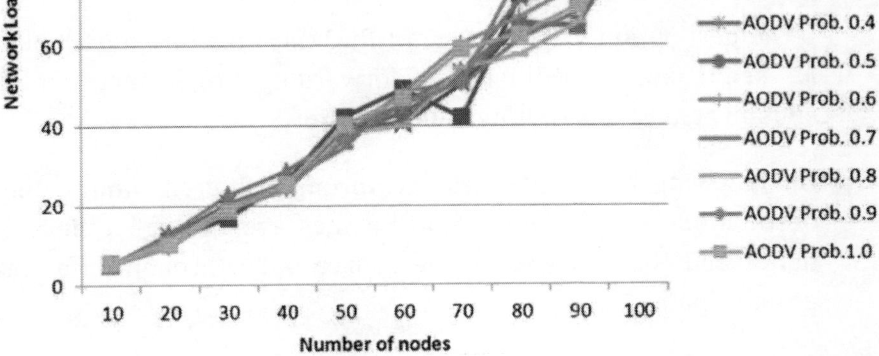

Figure 3. MAC LOAD AGAINST THE NUMBER OF NODES.

All the above results have been used by the authors to develop a technique to change dynamically the forwarding probability according to the network density. In the following section, a heuristic approach using the Bayesian theory is presented.

5. THE MODIFICATION OF AODV RREQ MECHANISM USING BAYESIAN PROBABILITY APPROACH

Below are the steps that are followed during the RREQ and RREP processing mechanism of AODV:

- When a source node, *S*, tries to send a packet to destination *D*.
- If *S* does not know the next hop for *D*, then it broadcasts a route request message.
- The RREQ message propagates in all directions to reach the destination *D*.
- All the intermediate nodes that receive the RREQ message forward the packet to all its one hop neighbours.
- If the destination, *D*, receives an RREQ message through a node *M*, then it sends an RREP to *S* by forwarding it to *M* since *M* may contain at least one routing table entry for *S*.
- On receiving the RREQ message through different nodes, the destination *D* will send the RREP message through different nodes and they may reach the source node through different possible paths.
- At the end, the source node *S* will have different possible resolved paths to select from based on defined criteria.

In this RREQ mechanism, we intervene and use a version of the modified Bayesian probability theory in order to improve the function of the RREQ. In this section, the Bayesian probability theory and similar research that have been conducted by other researchers, together with a detailed explanation of our proposed algorithm, will be presented.

6. BAYESIAN PROBABILITY THEORY

The Bayesian probability theory[25] manipulates conditional probabilities. The joint probability of two events, *A* and *B*, can be expressed as

$$\mathrm{P}(AB) = P(A/B) * P(B) \mathrm{or} P(B/A) * P(A) \tag{4}$$

In Bayesian probability theory, one of these events is called hypothesis and denoted with *H* and the other is called data and denoted with *D*. The Bayesian probability theory assesses the probability of the observed data *D* resulting from hypothesis *H*.

$$P(H/D) = \frac{P(D/H) * P(H)}{P(D)} \tag{5}$$

The term $P(H)$ is called the prior because it reflects prior knowledge before the data are considered. The term $P(D)$ is the result of the sum of $P(D|H)P(H)$ over all *H*. Finally, the term $P(H|D)$ is known as the posterior and reflects the probability of the hypothesis after consideration of the data.

There have been many efforts to us Bayesian probability in order to alleviate the flooding phenomenon. Jain et al.[26] proposed a heuristic algorithm with a route establishment technique using Bayesian approach. This algorithm improves the performance of route discovery by ameliorating the cost of route establishment using a history-based Bayesian method together with the relative region information of the destination node. The drawback of this effort is that it process and compares many information from each node (region, distance, status, destination node id and source node id) and which makes the algorithm complex, energy inefficient and time consuming in order to find the best route. Moreover, the simulations that have been conducted by other researchers were limited both in network size (number of nodes) and speed of the nodes. In large networks (bigger than 30 nodes), the behaviour of the network is different and there is bigger traffic congestion which causes unnecessary collision and packet overhead.

de Leoni et al.[27] have used the Bayesian probability theory in order to predict the disconnections that may occurred in MANETs. This research assumes that the nodes that are equipped with GPS receiver/transmitter and uses Bayesian filter to predict possible disconnections and control adjacent nodes to move to a location that that will provide continuous coverage. The disadvantage of the proposed algorithm is that it assumes that all nodes are equipped with GPS.

In our approach, we are using the Bayesian probability theory which is modified according to two major assumptions:

1. The posterior probability P_i is the forward probability of five nodes which is 1 (100%). In this case, when we have only five nodes or less the forward probability is 100%.

2. The forward probability scheme P_i depends on the minimum expected neighbours (*d*) and the number of neighbours (*n*). In this case, it multiplied the Bayesian probability P_i by ½ (*n*/*d*).

Using this approach the RREQ mechanism was modified to broadcast only to a percentage of nodes for discovery the route. An example of how the Bayesian technique is implemented and how it affects the AODV protocol is described in details.

Assume that a node D sends (forward) messages to anodes D_i then using the probability forwarding scheme P_i is

$$P(D/D_1) = \frac{P_{(D/D_1)} * \boldsymbol{P}_{(i)}}{(D_1/D) * \boldsymbol{P}_{(D_1)} + \boldsymbol{P}_{(D_2/D)} \boldsymbol{P} * \boldsymbol{P}_{(D_2)} + \cdots + \boldsymbol{P}_{(Dn/D)} * \boldsymbol{P}_{(D_n)}}$$

(6)

If it is assumed that there are nine nodes, in this case P_i is the posterior probability which is the forwarding probability of five nodes which is 1 [when we have only five nodes or less the forward probability is 1 (100%)]. The forward probability is calculating from the above equation:

$$P_{(D/D_i)} = \frac{\frac{1}{2} * 1}{\frac{1}{2} * \frac{1}{2} + \frac{1}{2} * \frac{1}{2} + \frac{1}{2} * \frac{1}{2} + \frac{1}{2} * \frac{1}{2} + \frac{1}{2} * \frac{1}{2} + \frac{1}{2} * \frac{1}{2} + \frac{1}{2} * \frac{1}{2} + \frac{1}{2} * \frac{1}{2} + \frac{1}{2} * \frac{1}{2}}$$
$$= \frac{\frac{1}{2} * 1}{\frac{9}{4}} = \frac{4}{18} = 0.22$$

(7)

This is the probability that if we forward the message with probability of 100% using the above approach 22% of the nodes will receive the message. This is for when $P(D1/D) = P(D1) = \frac{1}{2}$ which is an independent probability that explains that the probability of the node receiving the message is 50% probability.

Table3 presents the variables that have been used in the proposed algorithm.

Table 3. Definition Of Aodv_Ext_Bp Parameters

n	The total nodes in the network
F_i	Any node F_i, $i = 1,2,\ldots,n$ that receives the RREQ message
P_i	Packet forwarding probability derived from neighbour node count
d	Minimum node density or minimum expected neighbours—if the node density at a forwarding node F_i is less than this, then that node will forward the RREQ message to avoid the probability of path failure at that location. This means that the nodes will forward the packet without any further condition checking. In this study d is set to 5
r	Random number (between 0 and 1). This is used to generate varying conditions in the network. By this variable we are trying to avoid flooding in the RREQ mechanism as it compares each time to P_i which must be smaller than 1 (100%)

If the RREQ is received from an intermediate node then there will be at least one possible route which includes that node in its route list. Therefore, if only selected nodes are allowed to forward the RREQ packet, then only these nodes will be included in the path list. In this proposed scheme, the neighbourhood density of an intermediate node is considered as a criterion in RREQ forwarding decision at intermediate node. It denotes that if the number of nodes in the neighbourhood is high, then the Bayesian probability of any node transmitting will decrease and therefore reduces the transmission overhead. Random selection of nodes from the neighbourhood set increases the chances of full network coverage. Greater savings could be achieved by using a range-dependent technique to select nodes for transmission but this can only be achieved at the cost of greater complexity.

7. PROPOSED AODV_EXT_BP ALGORITHM

➢ Any node F_i $i = 1,2,\ldots n$ receiving the RREQ message will process the packet as follows:

➢ If the RREQ message originated from the node S_i or received at destination node D_i just process it in the normal way. If F_i is not the S_i or D_i then it will be an intermediate node:

➢ The node F_i will resolve the neighbourhood density D_i The Bayesian Probability P_i is:

$$P_i = \frac{P(D|D_i) * P(i)}{P(D1|D) * P(D1) + P(D2|D) * P(D2) + \ldots + P(D_n|D) * P(D_n)} * \tfrac{1}{2}(n/d)$$

If $P_i < r$ then

Forward the RREQ message

Else

Ignore and drop the RREQ message

End

(Here r is a random number between 0 and 1)

The proposed algorithm is based on the Bayesian probability theory and disseminates the RREQ packet not at the whole network of nodes but only at those nodes that the algorithm chooses which is lesser than the complete network. In that case, the major differences are the following:

- One of the main advantages of the proposed algorithm is that reduces the number of re-broadcasts without significantly compromising on its reachability, while the traditional probability scheme that the AODV uses, produces more re-broadcasts.

- The Bayesian probability scheme forwards the RREQ packet into lessen nodes than the traditional probability scheme.
- .By using the Bayesian probability scheme the power consumption is less by 3.3% than the traditional probability scheme.

8. SIMULATION AND METRICS OF AODV RREQ MECHANISM USING THE BAYESIAN APPROACH

The NS2 has been used to evaluate the protocols[28–31]. The simulations were carried out to assess the performance of the routing protocols with network sizes of 10, 20, 30, 40, 50, 60, 70, 90 and 100 nodes with mobile node speeds between 1 and 30 m/s. The simulation scenario has been repeated ten times in order the procedure and the results to be reliable so all the values on each graph are the average of the ten times simulation scenarios. For simplicity, in all cases the nodes send CBR over UDP. The metrics that have been used to evaluate the performance of the network and protocols are the following[24]:

- *Number of packets dropped*: This is the number of data packets that are not successfully sent to its destination.
- *Consumed power:* The average consumed battery power.
- *Throughput:* This measures how well the network can constantly provide data to the sink. Throughput is the number of packet arriving at the sink per millisecond.
- *MAC Load:* This is the ratio of the number of MAC layer messages propagated by every node in the network to the number of data

packets successfully delivered to all destination nodes. In other words, the MAC Load is the average number of MAC messages generated for each data packet successfully delivered to the destination.

- *Control message overhead:* This control message overhead is the total routing control messages transmitted and received in the network.

The simulation configuration and specification used are specified in Table4.

9. RESULTS AND DISCUSSIONS

Figure4 shows that AODV_EXT_BP consumes less power than the other five protocols. Most importantly, the power consumption of AODV_EXT_BP-based network improves in comparison to that of already improved version of AODV the AODV_EXT and the standard AODV. The AODV_EXT_BP performances best in all the range of network densities. DSR protocol-based networks consumes more energy compared to AODV_EXT and AODV_EXT_BP but shows better performance in comparison to AODV when the number of nodes in the network is close to 100 nodes. On the contrary, the power consumption in networks using DSDV and OLSR rises steadily starting from fairly high levels. With increasing number of nodes, the energy depletion of OLSR-based networks increases faster than those for the other protocols. OLSR protocol uses a mechanism that constantly updates information about nodes in the neighbourhood and therefore consumes more energy. As the number of nodes in the network increases, more updates are required and hence proactive protocols perform poorly, especially when the network is subject to changes, e.g. in mobile environment.

Table 4. Simulation Parameters and Configuration Of AODV, AODV_EXT, AODV_EXT_BP, OLSR, DSDV, DSR

Routing protocols	**AODV, AODV_EXT_BP, AODV_BP, OLSR, DSDV, DSR**
Topographical area	800×800 m^2
Number of nodes	10, 20, 30, 40, 50, 60, 70, 80, 90, 100
Mobility	1–30 m/s
Channel type	Wireless channel
Radio-propagation model	TwoRayGround
Network interface type	WirelessPhy
MAC type	802_11
Interface queue type	DropTail/PriQueue
Antenna model	OmniAntenna
Max packet in Queue	50
Transport/traffic type	CBR over UDP
TxPower of the nodes	0.1819 W
RxPower of the nodes	0.0501 W
IdlePower of the nodes	0.0350 W
Initial energy of the nodes	1000.0 J
Scenario simulation repetitions	10

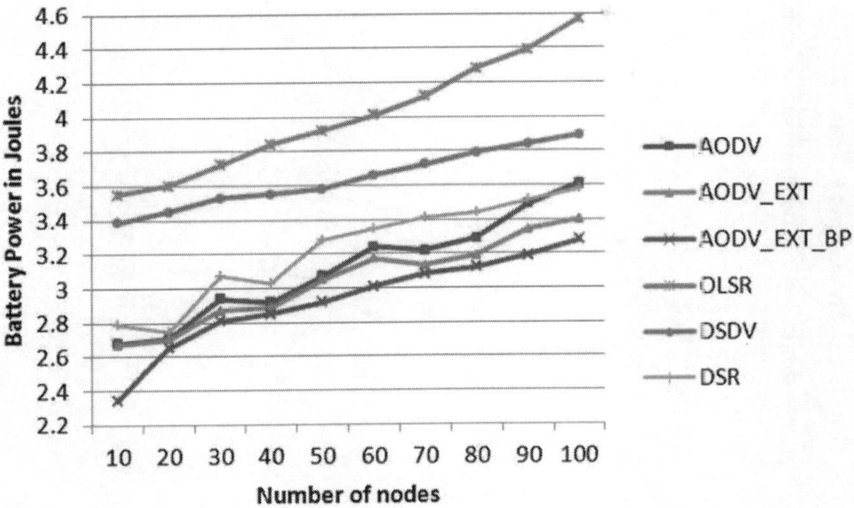

Figure 4. The Average Consumed Power.

Figure5 shows the performance of the protocols based on data throughput. It shows that the traditional DSR achieves the best performance. However, AODV_EXT_BP shows very good performance in average density networks (until 50 nodes). With AODV_EXT_BP, AODV_EXT and AODV protocols, every node does not need to keep information regarding the route between two nodes. This reduces the amount of signalling required for route discovery and maintenance. OLSR and DSDV both show poor performances compared to the four protocols. This is because both are proactive protocols and require table updates and generate relatively high messaging overhead that can cause congestion in large networks, especially in mobile networks, and reduces data rate performance of the network. However, these protocols are better suited to low data rate transmission because their self-updating scheme ensures connectivity rather than the availability of bandwidth for application data. But in our simulation results the AODV_EXT_BP performs, also, as well as the other two routing protocols (OLSR, DSDV).

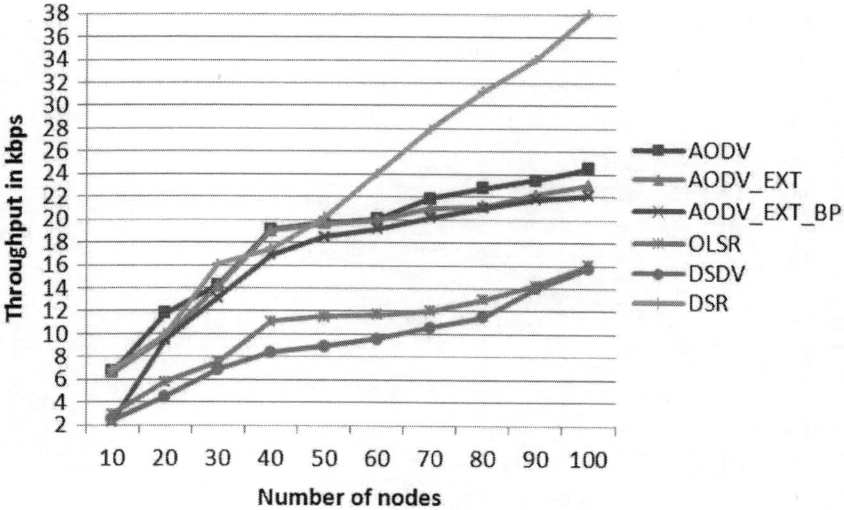

Figure 5. Comparison Of Data Throughput For Various Network Sizes (Nodes).

Figure6 shows MAC loading of the protocols. It shows that for large networks, a relatively high number of messages are generated by OLSR and DSR-based networks. This explains why the number of successfully delivered data packets is very low and therefore increases re-transmissions. DSDV and AODV exhibit only moderate increases. The figure shows that the use of density-based scheme as applied in AODV_EXT_BP significantly reduces the number of routing messages in the network. The proposed scheme reduces the amount of messages retransmitted in every range of the network density; the improvement is 50% better than the standard AODV and by 3% over the AODV_EXT.

Figure7 shows the routing control message overheads (routing load). In the case of AODV_EXT_BP, it is lower than that of the improved version of standard AODV, the AODV_EXT, and the original AODV protocol. DSR and DSDV show better performances due to the fact that they transmit and receive the least number of control messages. On the other hand, OLSR protocol has the worst performance of all the protocols and this degrades significantly as the number of nodes is between 30 and 40.

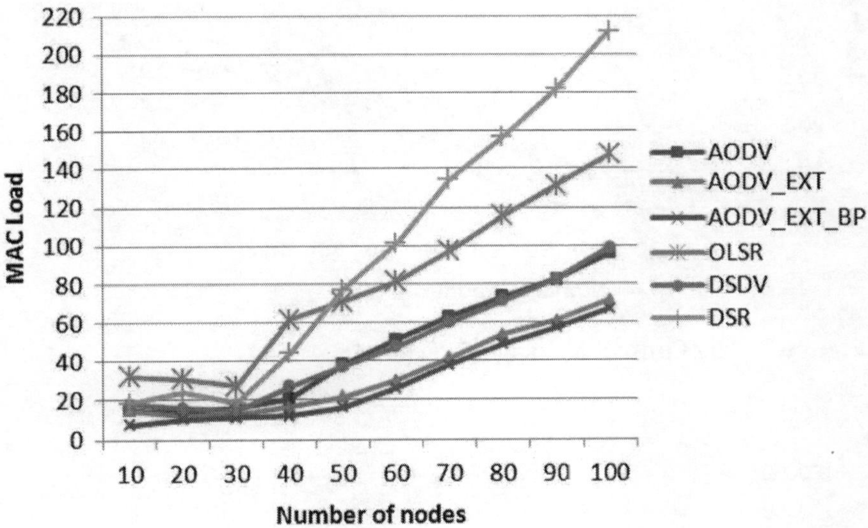

Figure 6. Mac Load Against The Number Of Nodes.

Figure8 shows the dropped packets in terms of network size. In case of AODV_EXT_BP, it is lower than that of the AODV_EXT and the original AODV protocol. DSDV presents almost the same results as AODV_EXT_BP because the mechanism that uses only adjacent nodes to send/receive packets has better performance in every scale of the network. OLSR and DSR show better performance in small-scale networks (until 30 nodes) after which both of them present low performances. The original AODV protocol shows very poor performance of all the protocols when the number of nodes exceeds 20.

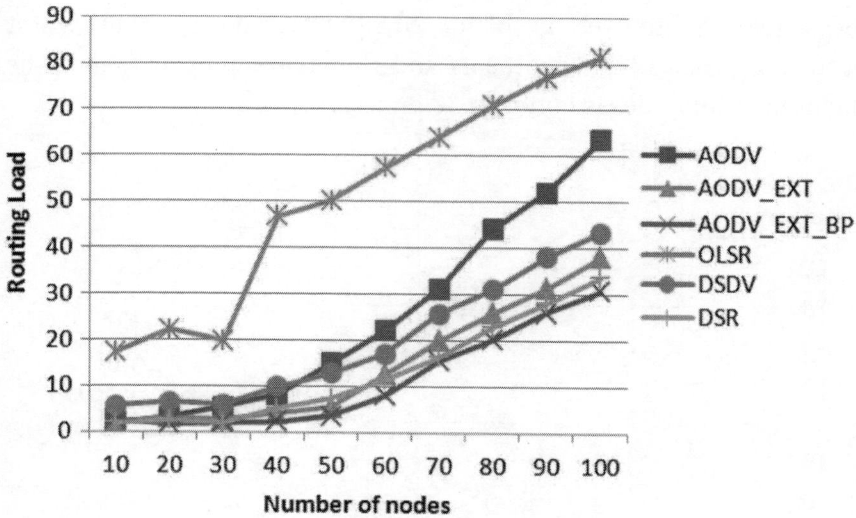

Figure 7. The Control Message Overhead.

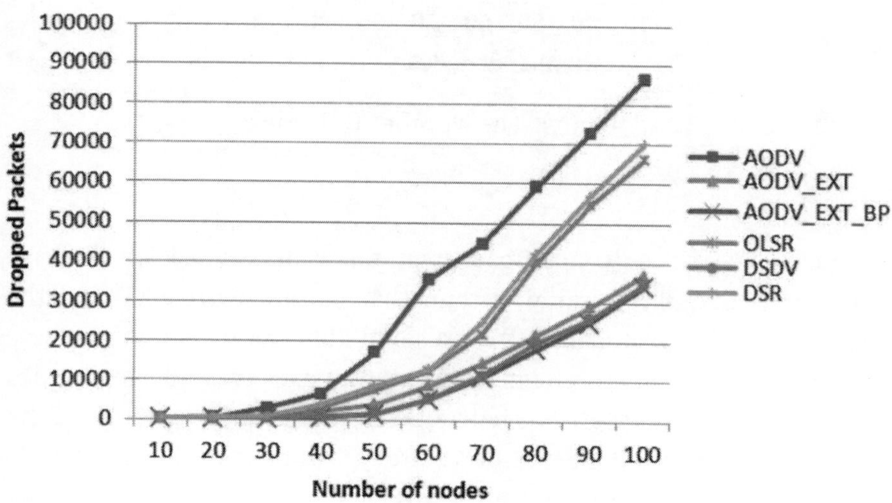

Figure 8. Dropped Packets In Terms Of Network Size.

10. CONCLUSIONS

The evaluation of six widely used protocols (AODV_EXT, AODV_EXT_BP, AODV, DSDV, DSR and OLSR) has been presented in this article. Their performances in different size networks and in mobile scenarios have been studied using simulations developed in NS2. AODV has been modified to use a modified Bayesian probabilistic approach for transmitting RREQ. The modified version has been named AODV_EXT_BP. Unlike in some probability-based approaches, where every node is assigned a fixed probability that does not ensure full network coverage, the technique proposed in this article combines concepts from maximum range node selection with node pruning to reduce redundant re-transmissions in route request but offer connectivity and better network coverage guarantees inherent in deterministic techniques.

The reduction in route request transmissions in a network using AODV_EXT_BP has resulted in 3.3% which is 0.3% better energy efficiency savings compared to AODV_EXT[12], more than 70% reduction in the number of dropped packets because of reduced packet collision and increased data throughput. The results from this simulation can be compared with those in the research conducted by Khelifa and Maaza[32] proposed the Energy Reversed Ad-Hoc On-Demand Distance Vector (ER-AODV) routing protocol and consumes up to 2.0% more power than AODV_EXT_BP. Moreover, AODV_EXT_BP improves the data throughput by more than 20% compared to the standard AODV and 12% more than ER-AODV. The results also show that proactive protocols, whilst they are more reliable in terms of connectivity, exhibit poor performance in large networks. Reactive protocols, on the other hand, are more suited to large networks. Both classes of protocols perform poorly in large mobile networks due to large overheads associated with routing as the nodes move. A hybrid protocol such as AODV offers a compromise and the technique proposed in this article to reduce redundant re-transmissions based on transmitting node neighbourhood density has produced very promising results when compared to standard protocols. This research has showed that fine tuning of protocols to suit specific applications or traffic scenarios to achieve optimum performance in ad-hoc networks is essential.

In the future, a different probability approach such as a Monte Carlo algorithm implementation at the route request discovery mechanism, it could be challenging research, investigates whether it alleviates the storm phenomenon or not.

COMPETING INTERESTS

The authors declare that they have no competing interests.

REFERENCES

1. Min R, Chandrakasan A: Top five myths about the energy consumption of wireless communication. Mob. Comput. Commun. Rev. 2003, 7(1):65-67. 10.1145/881978.881998

2. Kanakaris V, Ndzi D, Azzi D: Ad-hoc networks energy consumption: a review of the ad-hoc routing protocols. J. Eng. Sci. Technol. Rev. 2010, 3(1):162-167.

3. Ramdhany R, Grace P, Coulson G, Hutchison D: Dynamic deployment and reconfiguration of ad-hoc routing protocols. J. Internet Services Appl. 2010, 1(2):135-152. 10.1007/s13174-010-0010-y

4. Xu Y, Heidemann J, Estrin D: Geography-informed energy conservation for ad hoc routing. In Proceedings of the Seventh

Annual ACM/IEEE International Conference on Mobile Computing and Networking, July 16–21. (Rome) Italy; 2001:70-84.

5. Rabaey J, Ammer J, da Silva JL Jr, Patel D: PicoRadio: ad-hoc wireless networking of ubiquitous low-energy sensor/monitor nodes. In Proceedings of the IEEE Computer Society Annual Workshop on VLSI (WVLSI'00), April. Orlando, FL. USA; 2000:9-12.

6. Raghunathan V, Pereira CL, Srivastava MB, Gupta RK: Energy aware wireless systems with adaptive power-fidelity tradeoffs. IEEE Trans. Very Large Scale Integr. (VLSI) Syst 2005, 13(2):211-225.

7. Shen CC, Srisathapornphat C, Liu R, Huang Z, Jaikaeo C, Lloyd E: CLTC: a cluster-based topology control framework for ad hoc networks. IEEE Trans. Mob. Comput. 2004, 3(1):18-32. 10.1109/TMC.2004.1261814

8. Bani Yassein M, Bani Khalaf M, Al-Dubai A: A new probabilistic broadcasting scheme for mobile ad hoc on-demand distance vector (AODV) routed networks. J. Supercomput 2010, 53(1):196-211. 10.1007/s11227-010-0408-0

9. Bani Yassein M, Ould-Khaoua M, Mackenzei LM, Papanastasiou S: Performance analysis of adjusted probabilistic broadcasting in mobile ad hoc networks. Int. J. Wirel. Inf. Netw 2006, 13(2):127-140. 10.1007/s10776-006-0027-0

10. Sasson Y, Cavin D, Schiper A: Probabilistic broadcast for flooding in wireless mobile ad hoc networks. In Proceedings of IEEE wireless communication and networking conference (WCNC 2003), (2),March. New Orleans, LA, USA; 2003:1124-1130.

11. Zhang Q, Agrawal DP: Dynamic probabilistic broadcasting in MANETs. J. Parallel Distrib. Comput. 2005, 65(2):220-233. 10.1016/j.jpdc.2004.09.006

12. Kanakaris V, Ndzi DL, Ovaliadis K: Improving AODV performance using dynamic density driven route request forwarding. Int. J. Wirel. Mob. Netw. (IJWMN) 2011, 3(3):16-27. 10.5121/ijwmn.2011.3302

13. Survey: broadcasting in wireless sensor network, ad-hoc network and delay-tolerant network 2011.http://www.cs.tamu.edu/academics/tr/2011-9-2

14. Karthikeyan N, Palanisamy V, Duraiswamy K: Optimum density based model for probabilistic flooding protocol in mobile ad-hoc network. Eur. J. Sci. Res. 2010, 39(4):577-588.

15. Wu J, Dai F: Efficient broadcasting with guaranteed coverage in mobile ad-hoc networks. IEEE Trans. Mob. Comput. 2005, 4(3):259-270.

16. Jamal A: Probabilistic route discovery for wireless mobile ad-hoc networks (MANETs), PhD thesis. University of Glasgow; 2009.

17. Requena JC, Vadar T, Kantola R, Beijar N: AODV-OLSR scalable ad hoc routing proposal, in Symposium on Wireless Pervasive Computing. Espoo, Finland; 2006:1-5. Jan

18. Abd Rahman AH, Zukarnain ZA: Performance comparison of AODV, DSDV and I-DSDV routing protocols in mobile ad-hoc networks. Eur. J. Sci. Res 2009, 31(4):556-576.

19. The dynamic source routing protocol for mobile ad-hoc networks , Accessed 2 April 2008 http://tools.ietf.org/html/draft-ietf-manet-dsr-10

20. De Rango F, Fotino M, Marano S: EE-OLSR: energy efficient OLSR routing protocol for mobile ad-hoc networks. IEEE Military Communications Conference 2008, 1-7.

21. Qayyum A, Viennot L, Laouiti A: Multipoint relaying for flooding broadcast messages in mobile wireless networks. 35th Annual Hawaii International Conference on System Sciences (HICSS'02), Hawai; 2002:3866-3875. (9),Jan

22. Ad-hoc on-demand distance vector (AODV) routing 2000.http:// tools.ietf.org/html/draft-ietf-manet-aodv-08

23. Yoon J, Liu M, Noble B: Random waypoint considered harmful. In Twenty-Second Annual Joint Conference of the IEEE Computer and Communications (INFOCOM 2003). Ann Arbor, MI, USA; 2003:1312-1321. (2), March

24. Rahman A, Islam S, Talevski A: Performance measurement of various routing protocol in ad-Hoc network, Proceedings of the International MultiConference of Engineers and Computer Scientists 2009 (IMECS), (1). Hong-Kong; 2009:321-323.

25. Bayesian probability 2010.http://people.cs.kuleuven.be/~danny.deschreye/urks-slides2007-1x3.pdf , Accessed 5 Jan

26. Jain R, Parameswaran M, Hota C: An efficient on-demand routing protocol for MANETs using Bayesian approach. In Third International Conference on Communication Systems and Networks (COMSNETS), Jan. Bangalore, India; 2011:1-4.

27. de Leoni M, Humayoun S, Mecella M, Russo R: A Bayesian approach for disconnection management in mobile ad hoc networks. In 16th IEEE International Workshops on Enabling Technologies: Infrastructure for Collaborative Enterprises (WETICE 2007), June. Paris France; 2007.

28. Ns-2 network simulator , Accessed 2 May 2008 http://www.isi.edu/nsnam/ns/
29. CMU Monarch BPensions to NS-2 , Accessed 2 May 2008 http://www.monarch.cs.cmu.edu/cmu-ns.html
30. Marc Greis' Tutorial for the UCB/LBNL/VINT Network Simulator "ns" , Accessed 2 May 2008 http://web.uct.ac.za/depts/commnetwork/tutorial_ns_full.pdf
31. Installing OLSR on NS-2 , Accessed 20 May 2008 http://masimum.inf.um.es/um-olsr/html
32. Khelifa S, Maaza M: An energy multi-path AODV routing protocol in ad hoc mobile networks. In 5th International Symposium on I/V Communications and Mobile Network (ISVC), Oct. Oran Algeria; 2010:1-4.

CHAPTER 9

Some One Parameter Models for Continuous Random Variables Defined on the Interval [0, 1]

John J. Wiorkowski

The University of Texas at Dallas, Richardson, USA

ABSTRACT

The Beta Distribution is almost exclusively used for situations, after range normalization, wherein a continuous random variable is defined on the closed range [0, 1]. Since the beta distribution is intrinsically a two parameter distribution, this creates problems in some applications where specification of more than one parameter is difficult. In this note, two new classes of single parameter continuous probability distributions on a closed interval are introduced. These distributions remove some of the theoretical and practical problems of using the Beta Distribution for applications. The Burr Type XI Distribution has desirable characteristics for many applications especially when there is ambiguity in the definition of the specified parameter.

KEYWORDS

Probability Distribution; Risk Assessment; Reliability Assessment; PERT CPM

1. INTRODUCTION

The standard two parameter Beta Distribution is the most widely used distribution for situations wherein a continuous random variable is confined to a bounded interval $[a,b]$. After appropriate normalization to the interval $[0,1]$, the Beta Distribution provides a flexible family of probability density functions capable of modeling a wide variety of natural phenomena. There are situations, however, where the Beta Distribution cannot model natural phenomena or its use is problematic. In their recent book, Kotz and van Dorp [1] introduce and discuss the properties of other continuous families of distributions with bounded support. This widely extends the types of natural phenomena which can be modeled. But even this wide array of probability distributions have problems when modeling a low risk event. Consider the problem of providing a distribution for the situation where a subject matter expert subjectively estimates that the probability of a risky event as 0.01. Since the Beta Distribution is a two parameter model, this single estimate is inadequate for determining an appropriate Beta Distribution model. Except for the Triangular Distribution, all the alternative models in Kotz and van Dorp [1] are also at least two parameter models and thus indeterminate based on a single estimate of the risk probability. A substantial literature exists to aid the statistician in eliciting further information from subject matter experts to remove this indeterminacy (see O'Hagan et al. [2] for a review of the area). These techniques involve specifying a numerical estimate for another characteristic of the distribution such as the mean or a percentile. Taking the initial estimate as the mode and the estimate of some other distribution characteristic, allows a two parameter model to be fit (See Donaldson [3], Johnson [4], Lau, A. et al. [5], Lau H. et al. [6], Mohan et al. [7] and Premachandra [8]. Another example involves the management of large scale complex projects. A common methodology used in this situation is called PERT-CPM (Program and Evaluation and Review Technique-Critical Path Method). This technique has been used for the development of the Polaris Missile System and also for planning and managing the hosting activities of the International Olympics. In this approach the large project is broken down into a myriad of component parallel or sequential activities, each of which are uncertain in duration. Experts are asked to provide estimates of durations of these components. But as stated by Fazar [9] "PERT quantifies knowledge about the uncertainties involved in developmental programs requiring effort at the edge of, or beyond, current knowledge of the subject—effort for which little or no previous experience exists". In discussion with individuals involved in either of these two situations, I have found that although they are comfortable in providing a subjective "most likely" estimate of an event, most would have difficulty specifying any other characteristic of the distribution. Indeed many seem to hold that their subjective estimate is not only the most likely, (corresponding to the mode of the distribution) but simultaneously would be willing to give even odds that the actual probability is above or below their estimate (corresponding to the median). This ambiguity has also been reported by Trout [10]. Accordingly, there is a need for one parameter probability models to handle the real situation of having only one reliable estimate from a subject matter expert and for which the mode and median are very close. Of course it is possible to create probability distributions where the mode and median exactly coincide, however, there may be circumstances where other criteria may be paramount so that general methods for creating one parameter distributions may be of value.

2. TILTED DISTRIBUTIONS

Exponential tilting is a well known method (See Davison [11] Section 5.2 for example) which can be used to induce a one parameter family of distributions. Let $f(x)$ denote a probability density function on the closed interval $[0,1]$ with the property that $f(x) \geq 0$ and such that,

$$\int_0^1 f(x)\mathrm{d}x = 1 \qquad (1).$$

Define the moment generation function of $f(x)$ as,

$$\varphi_f(t) = \int_0^1 \mathrm{e}^{at} f(x)\mathrm{d}x \qquad (2)$$

then

$$g_f(x,t) = \frac{\mathrm{e}^{at} f(x)}{\varphi_f(t)} \qquad (3)$$

is also a probability density function on the closed interval $[0,1]$. If $f(x)$ has no parameters, then $g_f(x,t)$ defines a one parameter family on $[0,1]$ with parameter t, where t can range over the interval $-\infty$ to ∞. The probability density function $g_f(x,t)$ has several desirable properties. First, the moment generating function of $g_f(x,t)$ is given by $\varphi_f(s+t)/\varphi_f(t)$ so that the mean and variance of the density function are given by the equations

$$\mu_f(t)=\frac{\mathrm{d}\ln\varphi_f(t)}{\mathrm{d}t},\sigma_f^2(t)=\frac{\mathrm{d}^2\ln\varphi_f(t)}{\mathrm{d}t^2} \quad (4)$$

from which explicit formulae for the mean and variance can be obtained. The function $\ln\varphi_f(t)$ is the cumulant generating function of the density $f(x)$. Secondly, the form of (3) implies that $g_f(x,t)$ is in the exponential family of distributions. This family of distributions has been well studied and has several desirable properties. For example, in this case, if one has a random sample from $g_f(x,t)$, then one can obtain the maximum likelihood estimate of t by simply setting the sample mean equal to $\mu_f(t)$ and solving for t. The asymptotic variance of this estimate if given by $\left(n\sigma_f^2(t)\right)^{-1}$.

Finally, if $\ln f(x)$ is a concave function, then t can be uniquely related to the mode, λ, through the relationship,

$$t=-\frac{f'(\lambda)}{f(\lambda)} \quad (5)$$

where $f'(\lambda)$ is the first derivative of $f(x)$ with respect to x, evaluated at λ. This means that one can specify a distribution by specifying the mode λ, use (5) to find t, and then use (4) to determine the basic statistical properties of the distribution.

The choice of $f(x)$ is extremely broad, but as pointed out by one of the referees, it is best to start with $f(x)$ which is symmetric about the point $x = 0.5$ to guarantee that any modal value between 0 and 1 can be modeled As examples of the above, consider three simple choices for $f(x)$ on the closed interval $[0,1]$. The first is the Beta (2, 2) distribution with probability density function,

$$f_{\text{Beta}(2,2)}(x) = 6x(1-x) \tag{6}$$

The second is the Gilbert distribution (Edwards [12]) with probability density function,

$$f_{\text{Gilbert}}(x) = \frac{\pi \sin(\pi x)}{2} \tag{7}$$

The third is a translated version of the Raised Cosine distribution (Proakis [13], p. 189), and is also a special case of the Burr Type XI distribution (Burr [14] and Kotz and Johnson [15], p. 335). It has probability density function,

$$f_{\text{Burr}}(x) = 1 - \cos(2\pi x) \tag{8}$$

These probability density functions are shown in Figure 1. As can be seen, the Beta (2, 2) and Gilbert densities are quite close to one another. Further, the Burr density differs distinctly from the other two being much more concentrated around the value 0.5, and showing much greater tapering as x approaches 0 or 1. (As pointed out by the referee, in some applications one might prefer to start with symmetric beta distributions but the above three choices illustrate the distribution construction approach adequately.)

Figure 2 shows the three tilted distributions with a mode, λ, of 0.01. The values of t corresponding to a mode of 0.01 are –98.9898 for the tilted beta, –99.9671 for the tilted Gilbert, and –199.9342 for the tilted Burr distribution.

As can seen, the tilted Beta and Gilbert densities are almost indistinguishable and quite distinct from the tilted Burr density. This similarity between the tilted Gilbert and tilted Beta distributions seems to be typical for both low and high values of the mode λ.

The skewness of the distributions might be viewed as a desirable characteristic if one believed that experts tend to underestimate the probability of rare events. (In a reliability situation with a mode say of 0.99, the tilted densities would be left skewed and model a situation wherein experts tend to overestimate the reliability.)

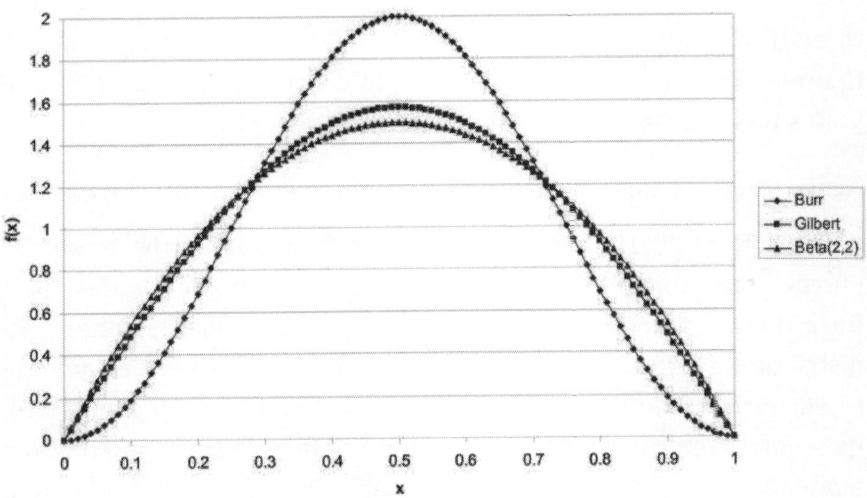

Figure 1. Probability density functions.

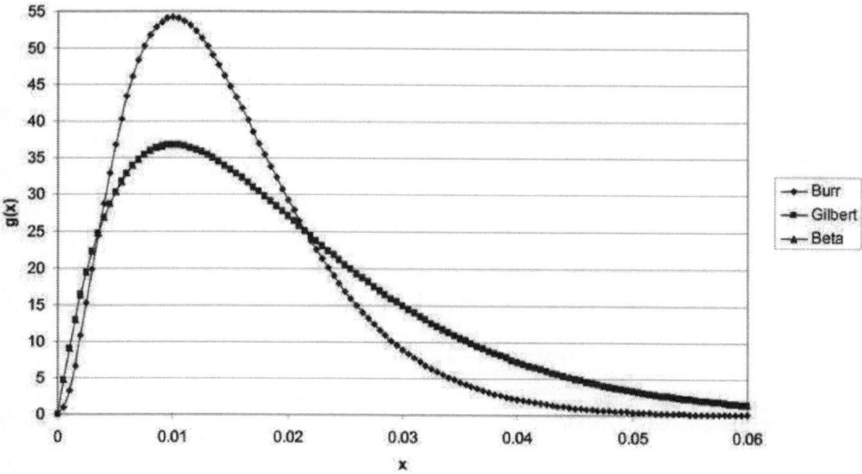

Figure 2. Tilted density functions mode = 0.01.

Figure 3 shows the relationship between λ and $\text{Prob}_f(X \le \lambda)$ for the three tilted distributions. The closer $\text{Prob}_f(X \le \lambda)$ is to 0.5, the closer the median and mode of the distributions are (Berny [16] uses one minus this probability as a parameter in his model).

Again, there is almost no difference between the tilted Beta and tilted Gilbert distributions over the whole range of the mode λ. Further the probabilities of being less than the mode for the tilted Burr distribution are uniformly closer to 0.5 than the other two distributions. However, the deviation from 0.5 is large even for the tilted Burr distribution. Accordingly, if one agreed with Trout [10], none of these distributions would provide Adequate models for modeling expert assessment of probabilities since the modes and medians do not seem close.

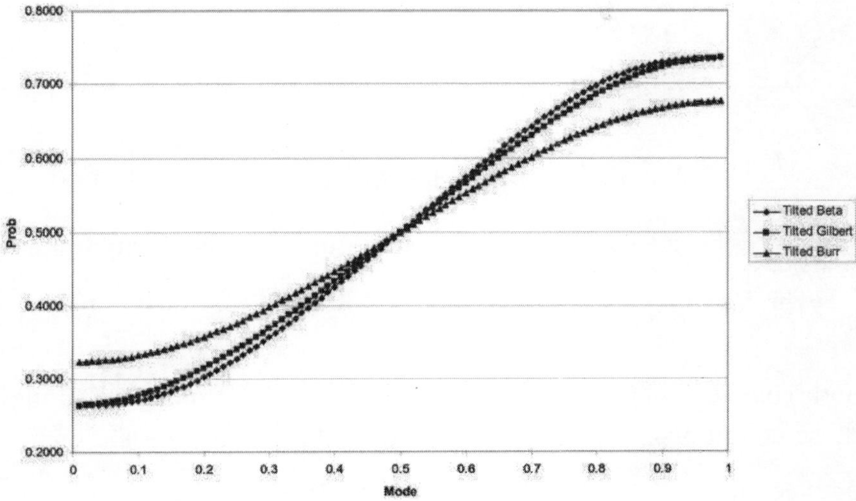

Figure 3. Probability of being less than the mode.

3. MAX DISTRIBUTIONS

Let $F(x)$ be the cumulative probability distribution function corresponding to the probability density function $f(x)$ on $[0,1]$, i.e.

$$F(x) = \int_0^x f(z)\,dz \tag{9}$$

Then for $K \geq 1$, $[F(x)]^K$ is also a cumulative probability distribution function on the range $[0,1]$. Indeed, if K is an integer, n, then $[F(x)]^n$ is the cumulative distribution function of the maximum value of X from a random sample of size n where X has probability density function $f(x)$. Define,

$$h_f(x,K) = Kf(x)F^{K-1}(x),$$
$$K \geq 1 \tag{10}$$

then $h_f(x,K)$ is a probability density function on $[0,1]$. Further, $h_f(1-x,K)$ is also a probability density function, the rotated image of $h_f(x,K)$ about the fixed point $x = 0.5$. If $\ln f(x)$ and $\ln F(x)$ are both concave, then $h_f(x,K)$ has a unique mode λ on $[0,1]$. Under these conditions, it is straightforward to show that the relationship between K and λ is given the by the equations,

$$K = \begin{cases} 1 - \dfrac{f'(\lambda)F(\lambda)}{f^2(\lambda)}, & \text{if } \lambda \geq 0.5 \\ 1 - \dfrac{f'(1-\lambda)F(1-\lambda)}{f^2(1-\lambda)}, & \text{if } \lambda \leq 0.5 \end{cases} \tag{11}$$

where $f'(x)$ is the first derivative of $f(x)$ with respect to x. Accordingly, one can generate a probability distribution by specifying the mode λ, use (11) to find K and if the mode is greater than 0.5 use (10) to find the density function. If the mode is less than 0.5, then one uses (10) with x replaced by $1-x$. Percentiles x_p for the density $h_f(x,K)$ can easily be found by solving the equation

$$x_p = F^{-1}\left(p^{1/K}\right) \tag{12}$$

Unfortunately, the form of (10) does not lend itself to closed form solutions for moments.

If one has a random sample of size n from $h_f(x, K)$, then the maximum likelihood of K is,

$$\hat{K} = \max\left(-\frac{n}{\sum_{i=1}^{n} \ln F(x_i)}, -\frac{n}{\sum_{i=1}^{n} \ln F(1-x_i)}\right) \tag{13}$$

with asymptotic variance K^2/n. A form of this distribution was discussed by Topp and Leone [17] and further discussed in Kotz and van Dorp [1].

Figure 4 shows the three MAX distributions for the case $\lambda = 0.01$. The BetaMAX distribution would have K = 1667, the GilbertMAX distribution would have K = 2026.59, and the BurrMAX distribution would have K = 101321.5.

As can be seen, as in the case of the Tilted distributions, the BetaMAX and GilbertMAX densities are indistinguishable and quite distinct from the BurrMAX distribution. Both the BetaMAX and GilbertMAX distributions show right skew (they would be left skewed if $\lambda > 0.5$) while the BurrMAX is almost symmetric. In contrast to the Tilted distributions shown in Figure 2, the Max distributions are much less variable with almost all the probability mass confined to the range 0 to 0.03. This is a much smaller range than the Tilted Beta and Tilted Gilbert which essentially go from 0 to 0.06. Accordingly, if one expected greater underestimation of probabilities by experts, the Tilted distributions would be preferred. However, as mentioned earlier, it may be the case that when asked to determine "most likely" values, experts are estimating the median rather than the mode.

Accordingly, in Figure 5, the probability of being less than the median is plotted as a function of the mode for the three Max densities discussed in this paper.

Figure 5 is plotted on the same scale as Figure 3. It is immediately clear, that all of the Max distributions have modes which are much closer to the median. The median and mode are remarkably close for the BurrMAX distribution, with $\text{Prob}(x \le \lambda)$ differing from 0.5 by a maximum value of 0.0281 at approximately λ = 0.28 and 0.72. Accordingly, the Max distributions seem better suited for handling the ambiguity of expert estimation of the "most likely value". Specifically, the BurrMax distribution, for which the mode and median are extremely close over whole range of possible values, seems an useful one parameter distribution for applications wherein individuals are asked to provide "most likely values".

4. THE BURRMAX OR BURR TYPE XI DISTRIBUTION

When $f(x)$ is the Burr distribution, (11) can be written as

$$K = \begin{cases} 1 + \dfrac{(\sin 2\pi\lambda - 2\pi\lambda)\sin 2\pi\lambda}{(1-\cos 2\pi\lambda)^2}, & \text{if } \lambda \ge 0.5 \\ 1 + \dfrac{(2\pi(1-\lambda) + \sin 2\pi\lambda)\sin 2\pi\lambda}{(1-\cos 2\pi\lambda)^2}, & \text{if } \lambda \le 0.5 \end{cases} \quad (14)$$

With the appropriate value of K, the cumulative distribution function of the Burr Type XI distribution is given by

$$H_{\text{Burr}}(x,K)=\begin{cases}\left(x-\dfrac{\sin 2\pi x}{2\pi}\right)^{K}, & \text{if } \lambda\geq 0.5\\ 1-\left(1-x+\dfrac{\sin 2\pi x}{2\pi}\right)^{K}, & \text{if } \lambda\leq 0.5\end{cases} \tag{15}$$

The first equation in (15) was first given by Burr [14], and given the designation Type XI by Kotz and Johnson [15] Vol. 1, p. 335, in their discussion of the Burr Family of distributions. The probability density functions of the BurrMAX distributions are

$$h_{\text{Burr}}(x,K) = \begin{cases}K\dfrac{(1-\cos 2\pi x)(2\pi x-\sin 2\pi x)^{K-1}}{(2\pi)^{K-1}}, & \text{if } \lambda\geq 0.5\\ K\dfrac{(1-\cos 2\pi x)(2\pi(1-x)+\sin 2\pi x)^{K-1}}{(2\pi)^{K-1}}, & \text{if } \lambda\leq 0.5\end{cases} \tag{16}$$

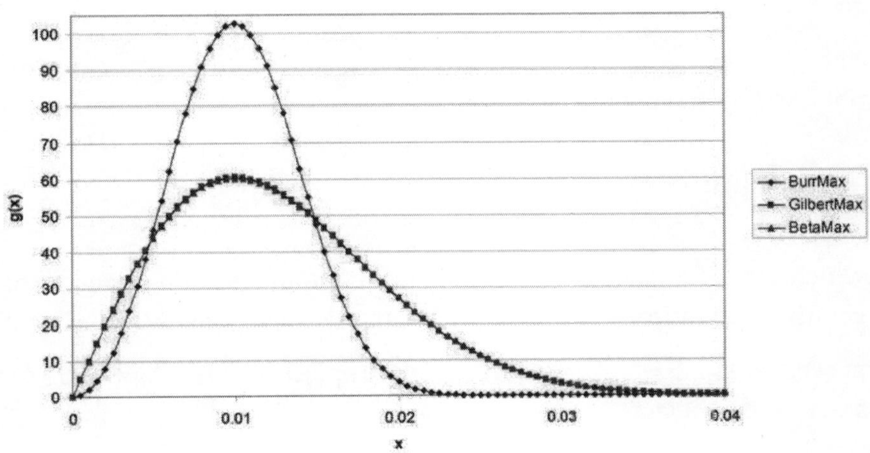

Figure 4. Max density functions mode = 0.01.

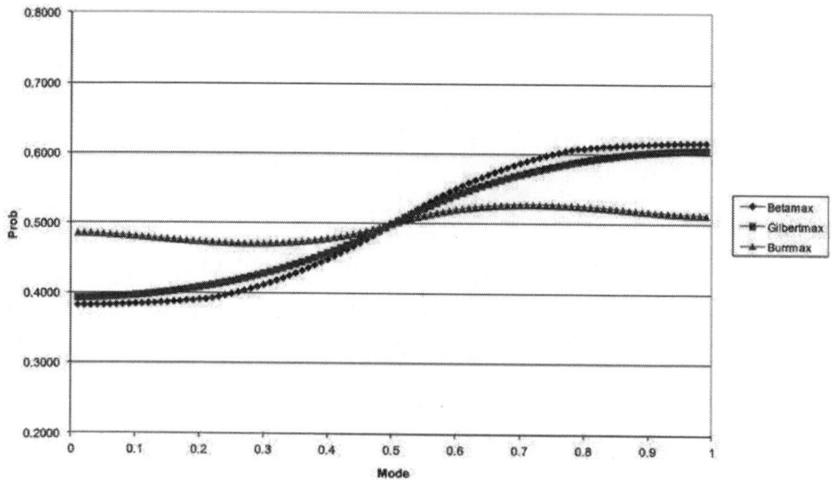

Figure 5. Probability of being less than the mode.

Since (14) are mixed trigonometric equations, there is no closed form equation to find the mode, λ, given K.

The median, $x_{0.5}$, or indeed any percentile x_p, for $\lambda \geq 0.5$, can be obtained by solving the equation

$$2\pi x_p - \sin\left(2\pi x_p\right) = 2\pi p^{1/K} \tag{17}$$

which, since it is a mixed trigonometric equation, has no closed form solution. Equation (17), however, can be solved using numerical procedures. If $\lambda \leq 0.5$, then to find x_p, replace p with $1-p$ in (54), solve, and subtract the solution from 1.

No closed form solution or general numerical solutions can be found for the mean and standard deviation of the Burr Type XI distribution. Accordingly, direct numerical integration of the integrals defining the first and second central moments was performed to obtain the resulting means and standard deviations as given in Table 1.

5. EXTREME VALUE DISTRIBUTION APPROXIMATION

In risk situations where one estimates very small probabilities, or in a reliability context when one estimates probabilities very close to 1, use of a Max distribution becomes problematic as the values of K become very large. One can capitalize, however, on the fact that for K = n, an integer, all of the Max distributions can be thought of as representing the distributions of the maximum (if $\lambda \geq 0.5$) or the minimum (if $\lambda \leq 0.5$) of n random samples taken from the appropriate distribution. Accordingly, one can use the theory of extreme values and extend the results from integer values, n, to a continuous value $K \geq 1$.

Following the discussion in Johnson et al. [18] Chapter 22, the distribution of the maximum of a sample of size n, for large enough n, for the distributions discussed in this paper, converge in law to a Weibull distribution which for $\lambda \leq 0.5$ has the cumulative distribution function is given by the equation

$$W(x,\lambda,\alpha)=\frac{1-\exp\left\{-\left(\frac{\alpha-1}{\alpha}\right)\left(\frac{x}{\lambda}\right)^{\alpha}\right\}}{1-\exp\left\{-\left(\frac{\alpha-1}{\alpha}\right)\left(\frac{1}{\lambda}\right)^{\alpha}\right\}} \tag{18}$$

If the value of α can be determined, then the above distribution becomes a function of λ alone and is a one parameter distribution. In Johnson et al. [18] Chapter 22, a method is given for finding α based on the cumulative distribution function $F(x)$ of the initial generating distribution. The result is that

$$\lim_{x\to 0}\frac{1-F(cx+1)}{1-F(x+1)}=c^{\alpha} \tag{19}$$

For the Burr distribution, (19) yields the value $\alpha=3$ so that for $\lambda\leq 0.3070$ or $\lambda\geq 0.6930$ the theory of Extreme Value Distributions indicates that the Burr Type XI distribution can be closely approximated to nine decimal places by a Weibull extreme value distribution on the range $0\leq x\leq 1$. Specifically, the cumulative distribution function is given by

$$W(x,\lambda)=\begin{cases}1-\exp\left[-\frac{2}{3}\left(\frac{x}{\lambda}\right)^3\right], \lambda\leq 0.307\\ \exp\left[-\frac{2}{3}\left(\frac{1-x}{1-\lambda}\right)^3\right], \lambda\geq 0.693\end{cases} \tag{20}$$

with corresponding probability density functions,

$$w(x,\lambda)=\begin{cases}\frac{2x^2}{\lambda^3}\exp\left[-\frac{2}{3}\left(\frac{x}{\lambda}\right)^3\right], & \lambda\leq 0.307\\ \frac{2(1-x)^2}{(1-\lambda)^3}\exp\left[-\frac{2}{3}\left(\frac{1-x}{1-\lambda}\right)^3\right], & \lambda\geq 0.693\end{cases} \tag{21}$$

Table 1. Characteristics of the Burr Type XI Distribution.

Mode λ	K	$E(x)$	Median	$Sd(x)$	$P(x \le \lambda)$	Mode λ	K	$E(x)$	Median	$Sd(x)$	$P(x \le \lambda)$
0.01	101321.5105	0.0102	0.0101	0.0037	0.4865	0.50	1.0000	0.5000	0.5000	0.1808	0.5000
0.02	12665.4686	0.0204	0.0203	0.0074	0.4863	0.51	1.0514	0.5081	0.5086	0.1785	0.5028
0.03	3752.9512	0.0307	0.0304	0.0111	0.4860	0.52	1.1071	0.5163	0.5174	0.1762	0.5054
0.04	1583.4525	0.0409	0.0405	0.0149	0.4856	0.53	1.1677	0.5247	0.5262	0.1737	0.5079
0.05	810.8730	0.0511	0.0507	0.0186	0.4852	0.54	1.2337	0.5332	0.5352	0.1712	0.5102
0.06	469.3777	0.0613	0.0608	0.0223	0.4846	0.55	1.3056	0.5419	0.5443	0.1685	0.5124
0.07	295.6901	0.0716	0.0709	0.0260	0.4840	0.56	1.3842	0.5507	0.5535	0.1657	0.5143
0.08	198.1811	0.0818	0.0810	0.0297	0.4833	0.57	1.4702	0.5597	0.5629	0.1629	0.5162
0.09	139.2700	0.0920	0.0912	0.0334	0.4826	0.58	1.5646	0.5688	0.5723	0.1599	0.5179
0.10	101.6001	0.1022	0.1013	0.0372	0.4819	0.59	1.6684	0.5781	0.5818	0.1569	0.5194
0.11	76.3986	0.1124	0.1114	0.0409	0.4811	0.60	1.7827	0.5875	0.5914	0.1537	0.5208
0.12	58.9053	0.1227	0.1216	0.0446	0.4803	0.61	1.9090	0.5970	0.6011	0.1505	0.5221
0.13	46.3842	0.1329	0.1317	0.0483	0.4796	0.62	2.0488	0.6066	0.6109	0.1472	0.5232
0.14	37.1870	0.1431	0.1418	0.0520	0.4788	0.63	2.2041	0.6163	0.6207	0.1438	0.5242
0.15	30.2797	0.1533	0.1520	0.0557	0.4780	0.64	2.3771	0.6262	0.6306	0.1403	0.5251
0.16	24.9916	0.1658	0.1637	0.0616	0.4773	0.65	2.5702	0.6361	0.6406	0.1367	0.5259
0.17	20.8745	0.1763	0.1740	0.0656	0.4765	0.66	2.7866	0.6461	0.6506	0.1331	0.5265
0.18	17.6214	0.1869	0.1844	0.0696	0.4758	0.67	3.0299	0.6562	0.6607	0.1294	0.5270
0.19	15.0169	0.1975	0.1948	0.0737	0.4752	0.68	3.3045	0.6664	0.6709	0.1256	0.5275
0.20	12.9069	0.2080	0.2052	0.0778	0.4746	0.69	3.6154	0.6766	0.6811	0.1218	0.5278
0.21	11.1794	0.2186	0.2156	0.0818	0.4740	0.70	3.9690	0.6869	0.6913	0.1180	0.5280
0.22	9.7514	0.2292	0.2260	0.0859	0.4735	0.71	4.3729	0.6973	0.7016	0.1140	0.5281
0.23	8.5606	0.2398	0.2363	0.0900	0.4730	0.72	4.8363	0.7077	0.7119	0.1101	0.5281
0.24	7.5598	0.2503	0.2467	0.0940	0.4726	0.73	5.3706	0.7181	0.7222	0.1061	0.5281
0.25	6.7124	0.2609	0.2571	0.0981	0.4723	0.74	5.9901	0.7286	0.7325	0.1021	0.5279
0.26	5.9901	0.2714	0.2675	0.1021	0.4721	0.75	6.7124	0.7391	0.7429	0.0981	0.5277
0.27	5.3706	0.2819	0.2778	0.1061	0.4719	0.76	7.5598	0.7497	0.7533	0.0940	0.5274
0.28	4.8363	0.2923	0.2881	0.1101	0.4719	0.77	8.5606	0.7602	0.7637	0.0900	0.5270
0.29	4.3729	0.3027	0.2984	0.1140	0.4719	0.78	9.7514	0.7708	0.7740	0.0859	0.5265
0.30	3.9690	0.3131	0.3087	0.1180	0.4720	0.79	11.1794	0.7814	0.7844	0.0818	0.5260
0.31	3.6154	0.3234	0.3189	0.1218	0.4722	0.80	12.9069	0.7920	0.7948	0.0778	0.5254
0.32	3.3045	0.3336	0.3291	0.1256	0.4725	0.81	15.0169	0.8025	0.8052	0.0737	0.5248
0.33	3.0299	0.3438	0.3393	0.1294	0.4730	0.82	17.6214	0.8131	0.8156	0.0696	0.5242
0.34	2.7866	0.3539	0.3494	0.1331	0.4735	0.83	20.8745	0.8237	0.8260	0.0656	0.5235
0.35	2.5702	0.3639	0.3594	0.1367	0.4741	0.84	24.9916	0.8342	0.8363	0.0616	0.5227
0.36	2.3771	0.3738	0.3694	0.1403	0.4749	0.85	30.2797	0.8467	0.8480	0.0557	0.5220
0.37	2.2041	0.3837	0.3793	0.1438	0.4758	0.86	37.1870	0.8569	0.8582	0.0520	0.5212
0.38	2.0488	0.3934	0.3891	0.1472	0.4768	0.87	46.3842	0.8671	0.8683	0.0483	0.5204
0.39	1.9090	0.4030	0.3989	0.1505	0.4779	0.88	58.9053	0.8773	0.8784	0.0446	0.5197
0.40	1.7827	0.4125	0.4086	0.1537	0.4792	0.89	76.3986	0.8876	0.8886	0.0409	0.5189
0.41	1.6684	0.4219	0.4182	0.1569	0.4806	0.90	101.6001	0.8978	0.8987	0.0372	0.5181
0.42	1.5646	0.4312	0.4277	0.1599	0.4821	0.91	139.2700	0.9080	0.9088	0.0334	0.5174
0.43	1.4702	0.4403	0.4371	0.1629	0.4838	0.92	198.1811	0.9182	0.9190	0.0297	0.5167
0.44	1.3842	0.4493	0.4465	0.1657	0.4857	0.93	295.6901	0.9284	0.9291	0.0260	0.5160
0.45	1.3056	0.4581	0.4557	0.1685	0.4876	0.94	469.3777	0.9387	0.9392	0.0223	0.5154
0.46	1.2337	0.4668	0.4648	0.1712	0.4898	0.95	810.8730	0.9489	0.9493	0.0186	0.5148
0.47	1.1677	0.4753	0.4738	0.1737	0.4921	0.96	1583.4525	0.9591	0.9595	0.0149	0.5144
0.48	1.1071	0.4837	0.4826	0.1762	0.4946	0.97	3752.9512	0.9693	0.9696	0.0111	0.5140
0.49	1.0514	0.4919	0.4914	0.1785	0.4972	0.98	12665.4686	0.9796	0.9797	0.0074	0.5137
0.50	1.0000	0.5000	0.5000	0.1808	0.5000	0.99	101321.5105	0.9898	0.9899	0.0037	0.5135

If $\lambda \le 0.307$, we have from the properties of the Weibull distribution [Johnson et al. [19] Chapter 21], that

$$E(x) = \lambda\left(\frac{3}{2}\right)^{\frac{1}{3}}\Gamma(1+1/3) \simeq 1.022206\lambda$$

$$\text{Median} = \lambda\left(\frac{3}{2}\ln(2)\right)^{\frac{1}{3}} \simeq 1.013069\lambda$$

$$\begin{aligned}\text{Var}(x) &= \lambda^2\left(\frac{3}{2}\right)^{\frac{2}{3}}\left[\Gamma(1+2/3) - \Gamma^2(1+1/3)\right]\\ &\simeq 0.138025\lambda^2\\ SD(x) &\simeq 0.371517\lambda\end{aligned} \tag{22}$$

For $\lambda \ge 0.693$, the corresponding results are,

$$\begin{aligned}E(x) &\simeq 1 - 1.022206(1-\lambda) \simeq 1.022206\lambda - 0.022206\\ \text{Median} &\simeq 1 - 1.013069(1-\lambda)\\ &\simeq 1.013069\lambda - 0.013069\\ \text{Var}(x) &\simeq 0.138025(1-\lambda)^2\\ Sd(x) &\simeq 0.371517(1-\lambda)\end{aligned} \tag{23}$$

For the BetaMAX and GilbetMAX distributions, (19) indicates that $\alpha = 2$ so that for $\lambda \le 0.1473$ or $\lambda \ge 0.8527$ both distributions can be closely approximated to nine decimal places by a Weibull extreme value distribution on the range $0 \le x \le 1$ which coincides with a form of the Rayleigh distribution Johnson et al. (1995), Chapter 18, Section [10].

Figure 6 shows the Extreme value distributions for $\lambda = 0.01$. As expected it's appearance is similar to that of Figure 4. The distribution for $\alpha = 3$ (limiting value of the Burr Type XI distribution) has expected value of 0.0102, a median of 0.0101, and a standard deviation of 0.0037. When $\alpha = 2$ (limiting form of the GilbertMax and BetaMax distributions) the corresponding values are 0.0125, 0.0118, and 0.0066.

6. APPLICATIONS

It is clear from the previous discussion that the Gilbert distribution and Beta (2, 2) distributions yield Tilted, MAX and Extreme Value distributions which are essentially numerically indistinguishable. Accordingly, I see no applications for any form of the Gilbert distribution. In risk or reliability studies, where λ would be expected to be close to 0 or close to 1 respectively, the Extreme value distributions would seem to be most useful. They have a relatively simple form and one can obtain good approximations to their moments using (20) and (21). Further one can obtain percentiles of the extreme value distributions using the formulas,

$$x_p = \begin{cases} \lambda\left[-\dfrac{\alpha}{\alpha-1}\ln(1-p)\right]^{1/\alpha}, \lambda \le 0.14 \\ (1-\lambda)\left[-\dfrac{\alpha}{\alpha-1}\ln p\right]^{1/\alpha}, \lambda \ge 0.86 \end{cases} \tag{24}$$

By replacing p or 1 – p in (24) with a uniformly distributed random variable U, one can easily simulate samples of any size from these distributions. The choice of whether to use $\alpha = 3$ or $\alpha = 2$ hinges on whether one believes that experts are truly estimating the mode and not the median. If they are, then choosing $\alpha = 2$ would seem to be preferred since it allows for expert under estimation of probabilities in the case of risk applications, or over estimation of probabilities in reliability applications. On the other hand, if there is ambiguity as to whether experts are estimating the mode or median when asked for the "most likely value", then using the Extreme Value distribution with $\alpha = 3$ would seem most appropriate since for this distribution the median and mode are almost identical.

In PERT or stochastic CPM applications where λ would not be expected to be either very small or very large, the Extreme Value distributions would not be appropriate. Typically either the Triangle distribution with mode λ or the Beta distribution with mean $(1+4\lambda)/6$ and standard deviation of $1/6$ have been used in this situation. The standard deviations of these two distributions as well as the BetaMax and Burr Type XI distributions are shown in Figure 7. Unlike the usual Beta approximation or Triangular distribution cases, the two Max distributions show the variability declining as the mode approaches 0 or 1.

It seems clear that the BetaMax and Burr Type XI distributions are better than both the usual Beta approximation model and Triangular distribution since for these two distributions the variability is substantially lower as one moves away from the middle of the modal range. If one was worried about the ambiguity of the term "most likely value", then one would use the Burr Type XI distribution instead of the Usual PERT model based on Figure 5 which shows the closeness of the median and mode of the Burr Type XI distribution. If this was not a concern then the choice would depend on one's conception of variability. However, if one wanted to keep variability at a reasonable level, again one is led to the Burr Type XI distribution with a standard deviation which is between 0.36 and 0.39 the value of the mode (when the mode is below 0.5).

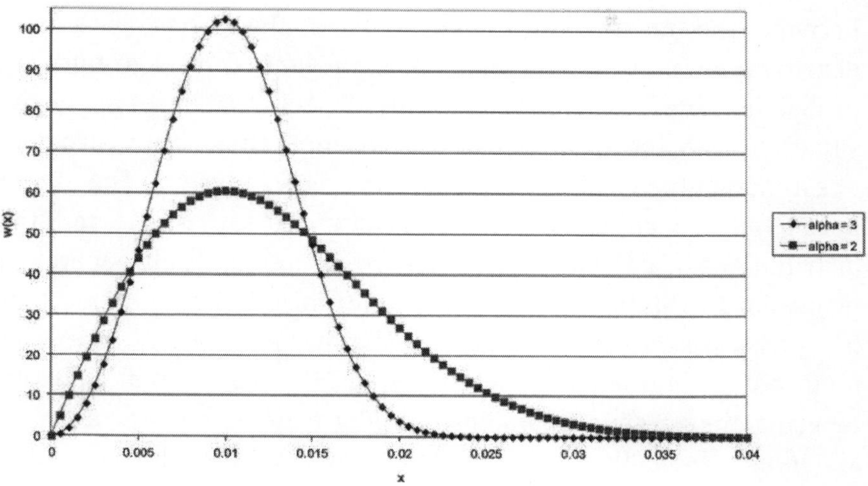

Figure 6. Weibull extreme value distributions mode = 0.01.

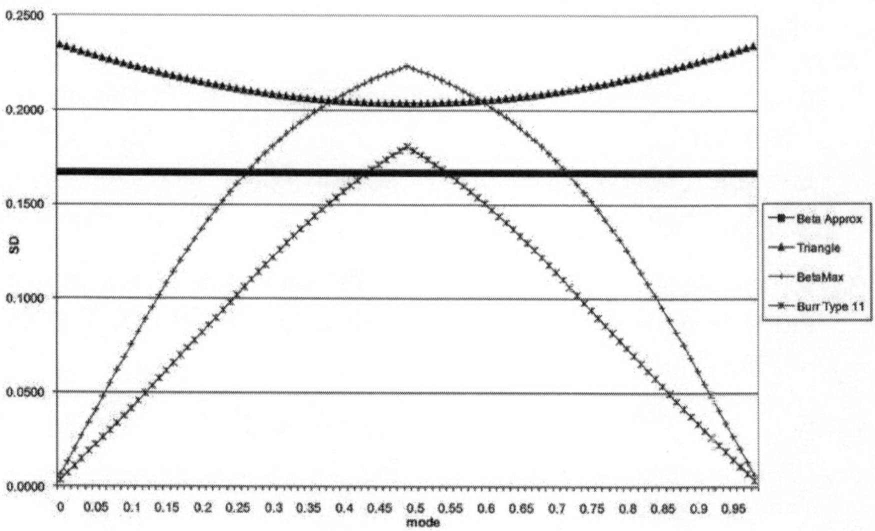

Figure 7. Standard deviations.

Perhaps the most useful application of these one parameter distributions is to allow experts with limited backgrounds in probability to more accurately specify their uncertainties about the situations they are working with. For example consider the problem of estimating the chances of a failure in a power system. The expert needs only to come up with one estimate, say 0.001, and the distributions discussed in this paper would automatically generate a plausible distribution for the uncertainty in this figure. Given that many risk assessment studies and complicated projects consist of hundreds to thousands of uncertain steps, the reduction in difficulty by using one parameter families should greatly ease the problem of assigning reasonable uncertainty to the myriad steps.

REFERENCES

1. S. Kotz and J. R. van Dorp, "Beyond Beta, Other Continuous Families of Distributions with Bounded Support and Applications," World Scientific Publishing Co., Singapore City, 2004.

2. A. O'Hagan, C. Buck, A. Daneshkhah, J. Eiser, P. Garthwaite, D. Jenkinson, J. Oakley and T. Rakow, "Uncertain Judgements, Eliciting Experts' Probabilities," John Wiley and Sons, Ltd., Chichester, 2006.

3. W. Donaldson, "The Estimation of the Mean and Variance of a PERT Activity Time," Operations Research, Vol. 13, No. 3, 1965, pp. 382-385.

4. D. Johnson, "The Triangular Distribution as a Proxy for the Beta Distribution in Risk Analysis," The Statistician, Vol. 46, No. 3, 1997, pp. 387-398.

5. A. Lau, H. Lau and Y. Zhang, "A Simple and Logical Alternative for Making PERT Time Estimates," IIE Transactions, Vol. 28, No. 3, 1996, pp. 183-192.

6. H. Lau, A. Lau and C. Ho, "Improved Moment-Estimation Formulas Using More than Three Subjective Fractiles," Management Science, Vol. 44, No. 3, 1998, pp. 346-351.

7. S. Mohan, M. Gopalakrishnan, H. Balasubramanian and A. Chandrashekar, "A Lognormal Approximation of Activity Duration in PERT Using Two Time Estimates," Journal of the Operational Research Society, Vol. 58, No. 6, 2007, pp. 827-831.

8. I. J. Premachandra, "An Approximation of the Activity Duration Distribution in PERT," Computers and Operations Research, Vol. 28, No. 5, 2001, pp. 443-452.

9. W. Fazar, "Program Evaluation and Review Technique," The American Statistician, Vol. 13, No. 1, 1959, p. 10.

10. M. Trout, "On the Generality of the PERT Average Time Formula," Decisions Sciences, Vol. 20, No. 2, 1989, pp. 410-412.

11. A. Davison, "Statistical Models," Cambridge University Press, London, 2003.

12. A. Edwards, "Gilbert's Sine Distribution," Teaching Statistics, Vol. 22, No. 3, 2000, pp. 70-71.

13. J. Proakis, "Digital Communications," 3rd Edition, McGraw Hill, Inc., New York, 1995.

14. I. Burr, "Cumulative Frequency Functions," Annals of Mathematical Statistics, Vol. 13, No. 2, 1942, pp. 215-232.

15. S. Kotz and N. Johnson, "Encyclopedia of Statistical Sciences," John Wiley and Sons, Inc., New York, 1982.

16. J. Berny, “A New Distribution Function for Risk Analysis,” Journal of the Operational Research Society, Vol. 40, No. 2, 1989, pp. 1121-1127.

17. C. W. Topp and F. C. Leone, “A Family of J-Shaped Distributions,” Journal of the American Statistical Association, Vol. 50, No. 269, 1955, pp. 209-219.

18. N. Johnson, S. Kotz and N. Balakrishnan, “Continuous Univariate Distributions,” Vol. 2, 2nd Edition, John Wiley and Sons, Inc., New York, 1995.

19. N. Johnson, S. Kotz and N. Balakrishnan, “Continuous Univariate Distributions,” Vol. 1, 2nd Edition, John Wiley and Sons, Inc. New York, 1994.

CHAPTER 10

A Probabilistic Method of Characterizing Transit Times for Quantum Particles in Non-Stationary States

Hae-Won Kim[1], Karl Sohlberg[2*]

[1]Department of Chemistry, Penn State Abington, Abington, PA, USA

[2]Department of Chemistry, Drexel University, Philadelphia, PA, USA

ABSTRACT

We present a probabilistic approach to characterizing the transit time for a quantum particle to flow between two spatially localized states. The time dependence is investigated by initializing the particle in one spatially localized "orbital" and following the time development of the corresponding non-stationary wavefunction of the time-independent Hamiltonian as the particle travels to a second orbital. We show how to calculate the probability that the particle, initially localized in one orbital, has reached a second orbital after a given elapsed time. To do so, discrete evaluations of the time-dependence of orbital occupancy, taken using a fixed time increment, are subjected to conditional probability analysis with the additional restriction of minimum flow rate. This approach yields transit-time probabilities that converge as the time increment used is decreased. The method is demonstrated on cases of two-state oscillations and shown to produce physically realistic results.

KEYWORDS

Conditional Probability; Non-Stationary Quantum Particle; Transit Time

1. INTRODUCTION

There are numerous situations where one might be interested in how quickly a quantum particle moves from one place to another. For example, electron transit times may influence such properties as binary switching speeds and alternating current conductance in molecular electronic devices. As another example, when a chemical reaction intermediate is tautomeric, the time required for the intermediate, prepared in one tautomer, to evolve into the other tautomer, may govern the overall reaction rate. In many cases this exchange of tautomers is essentially a proton transfer.

Investigating theoretically the time dependence of a quantum system begins with solving the time dependent Schrödinger equation (TDSE). Solving the TDSE yields the time dependent wavefunction

$\psi(t)$ and the corresponding probability density $\rho(t)=|\psi(t)|^2$. Consider the movement of a quantum particle between two spatial regions denoted A and B: From the time dependent probability density one may calculate the probability of the particle being in region A at any given time. One may also calculate the probability of the particle being in region B at any given time, but the time dependent wavefunction does not directly yield the time it takes for the particle to travel from A to B. In order to attain quantitative knowledge of the time-dependence of the flow of a quantum particle, it is essential to develop methods to extract this information from the time-dependent wavefunction $\psi(t)$ and/or the corresponding probability density $\rho(t)$. Owing to the non-commutativity of the quantum position and momentum operators, the time required for a quantum particle to get from one spatial location to another must be defined differently from the analogous quantity for a classical particle. Here we present an approach that expresses transit time in terms of probabilistic confidences.

To date, the great majority of theoretical treatments of quantum particle flow in molecular complexes have been of the time independent variety. Notable exceptions are Baer's treatment of anti-coherence in molecular electronics [1] and Pacheco and Iyengar's demonstration of how to extract the transmission probability from the time-dependence of electron density in donor-bridgeacceptor systems [2]. In related work, Guo has applied non-perturbative time-dependent quantum dynamics to study the phenomenon of superexchange in electron transfer [3], superexchange today being of central relevance in molecular transport junctions [4,5]. Coalson has investigated the accuracy of time-dependent perturbation theory methods to treat nuclear-electronic coupling in long-range electron transfer by benchmarking results for a model Hamiltonian to its exactly solvable limit [6]. Such long-range electron transfer is relevant in a wide variety of donor-bridge-acceptor systems [7]. These studies give insight into the time dependence of electron transfer, but none recover the transit time.

A special case from the class of problems encompassed by the question, “How fast does a quantum particle move from A to B?” is the question of tunneling time through a potential barrier. Tunneling time has been a matter of considerable discussion almost since the inception of quantum mechanics. Valuable reviews have been presented by Hauge and Støvneng [8] and more recently by Winful [9]. The latter review is especially valuable for its appendix, which addresses conceptual questions about tunneling. One issue of sustained interest, apparent superluminal transport across the tunneling barrier, now appears largely resolved [10], as it has been demonstrated that causality is not violated [11,12].

Although we consider a more general problem, the transit of a quantum particle from one spatial location to another, two “tunneling time” papers are more than tangentially relevant to the present work: First, Dumont and Marchioro [13] have given a clear explanation why the question, “How much time does a tunneling particle spend under the barrier?” is ill posed. The difficulty with this question is that its answer would require the simultaneous evaluation of observables corresponding to non-commutative operators. Dumont and Marchioro argue, as we alluded to above, that the question must be posed in a probabilistic manner. Second: Baskin and Sokolovskii [14] addressed tunneling time by generalizing the classical concept of transit time to the quantum mechanical case. We too have found it useful to reference a classical case, but rather than generalizing from a deterministic classical system, we draw analogy to a probabilistic classical system.

In this paper, we present a new way to characterize transit times by analyzing the time-dependent probability density. We then demonstrate the method on the two different systems that display two state oscillations. The movement of the quantum particle is investigated by following the time development of a localized wavefunction, which is expanded in a basis of eigenfunctions of the time independent Hamiltonian using standard methods. From the time development of the wavefunction, we show how to calculate the probability that the particle, initially localized in one orbital, has reached a second orbital after a given elapsed time; or conversely, how to calculate the time that must elapse for the particle to reach the second orbital with a given probability. Collectively, we refer to these questions as the "electron orbital-occupancy problem". A related question, "How fast can a quantum state change with time?" (under the control of a time-dependent Hamiltonian) is the subject of a work with that title by Pfeifer [15].

2. TWO STATE OSCILLATIONS

The time dependence of two-state oscillations is a topic from introductory quantum mechanics. We discuss such oscillations here principally to define the time-dependent probability function for a specific case to be analyzed here, but also to introduce the notation that will be used in the balance of the paper.

Consider an electron that oscillates between two spatially localized states (These localized states could represent the source and drain of a two-terminal molecular electronic device under no applied potential, or the donor and acceptor in a DBA system, but for purposes of this study we consider two generic spatially localized noneigenstates). We want to find out how long it takes for an electron that is initially prepared in one such state to reach a second state. Such a non-stationary state might be prepared with shaped laser pulses [16] or a sudden change in potential [17]. Here we will not concern ourselves with how the spatially localized state is prepared, but rather with how to analyze its time-evolution. To begin the problem we first solve the time-dependent Schrödinger equation (TDSE) by expanding the spatial wave function in a linear combination of eigenfunctions of the time-independent Schrodinger equation (TISE) according to canonical procedure.

Consider a time-independent Hamiltonian $\hat{H}$ for an electron in which there are two eigenstates φ_i that obey the relations:

$$\hat{H}\varphi_1 = E_1\varphi_1, \ \hat{H}\varphi_2 = E_2\varphi_2 . \qquad (1)$$

The TDSE for this system is:

$$i\hbar\partial\Psi/\partial t = \hat{H}\Psi . \qquad (2)$$

Given that $\hat{H}$ is time-independent, if we let: $\Psi = T(t)\Phi(r)$, then the system separates into time dependent and time independent parts with the solution of the time-dependent part being, $T(t) = \exp(-iEt/\hbar)$. Expanding the spatial wavefunction $\Phi(r)$ in a linear combination of eigenfunctions of $\hat{H}$, it follows that the general form of the solution to the TDSE is,

$$\Psi = T(t)\Phi(r) = \sum_k c_k \exp(-iE_k t/\hbar)\varphi_k , \qquad (3)$$

where the c_k are the expansion coefficients at $t = 0$. Since the basis functions are eigenstates, the probability of finding the system in a particular eigenstate at time t is, $\rho_k(t) = c_k \cdot c_k^*$, which is, in fact, independent of time. For our purposes, however, it is more useful to consider an expansion in non-eigenstate functions. For example, suppose that we describe a localized electron in a molecule as an expansion in atomic orbitals (AOs). We can follow the flow of charge by observing how the AO expansion coefficients vary as the time-dependent wavefunction evolves in time. Ultimately, we want to address the question; how long does it take for an electron that is initially localized in one AO to reach a second selected AO?

Suppose that the two eigenstates (φ_1 and φ_2) of the two-state system are expanded in a basis of two AOs; χ_1 and χ_2. In this basis, the TISE in secular form is,

$$\det\begin{vmatrix} \alpha - E & \beta - Es \\ \beta - Es & \alpha - E \end{vmatrix} = 0 \tag{4}$$

where $\alpha = \langle \chi_i | \hat{H} | \chi_i \rangle$, $\beta = \langle \chi_i | \hat{H} | \chi_j \rangle$, and $s = \langle \chi_i | \chi_j \rangle$. The solutions are,

$$E_1 = (\alpha + \beta)/(1 + s);\ \text{for which}\ \phi_1 = \chi_1 + \chi_2 \tag{5a}$$

$$E_2 = (\alpha - \beta)/(1 - s);\ \text{for which}\ \phi_2 = \chi_1 - \chi_2 \tag{5b}$$

If we use the approximation, $s = \delta_{ij}$, the roots are,

$$E = \alpha \pm \beta \tag{6}$$

Both eigenvalues are displaced from the site energy α by an amount dependent on the coupling matrix element β. Since α just sets the zero of potential, we can take the eigenenergies to be $E = \pm\beta$.

Suppose that we initially localize our electron by placing it exclusively in the AO χ_1. At $t = 0$ the spatial wavefunction, written as an expansion in AOs, is,

$$\Phi(r,t=0)=\chi_1=1\chi_1+0\chi_2 \qquad (7)$$

To write the time-dependent wavefunction, we must express this in the basis of eigenfunctions of $\hat{H}$. In other words, we must find the c_k in Equation (3). This is a standard change from an AO basis to a molecular orbital (MO) basis. From the definitions of ϕ_k in Equations (5a) and (5b), it follows that $c_1=0.5$ and $c_2=0.5$.

We can now write the time dependent wavefunction:

$$\begin{aligned}\Psi &= T(t)\Phi(r)\\ &= 0.5\exp(i\beta t/\hbar)\varphi_1+0.5\exp(-i\beta t/\hbar)\varphi_2\end{aligned} \qquad (8)$$

To follow the time evolution of the wavefunction in space, we need to transform from the MO basis back to the AO basis. Using the Euler relations, we obtain the time dependent AO-occupancy probabilities as:

$$\rho(\chi_1)=\cos^2(\beta t/\hbar) \qquad (9a)$$

$$\rho(\chi_2)=\sin^2(\beta t/\hbar) \qquad (9b)$$

As expected, $\rho(\chi_1)+\rho(\chi_2)=1$. Note that the frequency of oscillation between the two functions is $\beta/\hbar$. Note also that the oscillation frequency depends on the energy difference between the states, the greater the energy difference, the greater the frequency.

In oscillatory systems the period of oscillation (τ) bears a reciprocal relationship to the frequency $(\tau=2\pi/\omega)$ where (ω) is the angular frequency. As shown in the example above, a bound quantum system in a non-stationary state is characterized by such a frequency, (or frequencies in a system with more than two levels). The related period(s), however, is(are) a physically unsatisfactory metric of the lifetime of the associated spatially localized state. In Section 5 we demonstrate that this quantum exchange frequency suggests a rapid exchange between tautomers in cases when one would likely have to wait an aeon for one tautomer to evolve into another. The present paper addresses this dilemma.

Solution of the TDSE as outlined in this section yields the time-dependence of occupancy for each spatially localized orbital (AO). This standard procedure is easily generalized to an arbitrary number of AO basis functions. It is a standard result in quantum mechanics and answers the question, "what is the probability that a selected AO is occupied at a specific time t?" Herein we will demonstrate how to analyze the time dependent probability function to address our central question: Given that the electron is initially localized on one atomic orbital, how much time must elapse for the electron to reach the second orbital with a given probability? In the next section we look at an analogous classical system in order to develop the methods to answer this question. The system we look at is a "professor office-occupancy problem".

3. PROFESSOR OFFICE-OCCUPANCY PROBLEM

To develop the methodology, we first consider a classical problem analogous to the "electron orbital-occupancy problem": We ask the question; given a function that describes the probability of a professor being in her office as a function of time, what is the probability that the professor has occupied her office at least once during some elapsed time period? We term this the "professor office-occupancy problem". Exploration of this classical problem allows us to identify the variables that are important in the electron orbital-occupancy problem and to determine what assumptions might be necessary to produce a physically reasonable answer. Our approach is based on evaluating the time dependent probability density $\left(|\psi(t)|^2\right)$ at a fixed time increment. We find that conditional probability analysis is required to obtain a result that converges as the time increment is decreased. The understanding gained from studying the professor office-occupancy problem is then applied to solving the electron orbital-occupancy problem. In addition to studying the time dependence of electron orbital occupancy, the method is also demonstrated for proton exchange in a double well potential.

The professor office-occupancy problem is defined as follows: We take the probability of finding a professor in her office, $\rho(t)$, to be a sinusoidal function of time with a period of 24 hours and an amplitude of 0.79; a minimum at midnight at which there is 1% probability of finding her in her office; and a maximum of 80% probability at noon. The function is shown in Figure 1 (In this analysis we assume all days, whether weekend, weekday or holiday, are equivalent. A more complicated function with 7-day or 30-day periodicity could be used, but this would only serve to obfuscate the analysis without introducing any new physics). The probability function may be written as,

$$\rho(t) = 0.405 - 0.395 \times \cos(2\pi t/24), \quad (10)$$

where t is time in hours. The area under the curve, 9.72 hours/day, is the average amount of time the professor spends in her office each day.

This system can be thought of as a two-state oscillation. The professor is either "in" her office or "not in" her office, and she oscillates between these two "states" according to the assumed probability function (10). Now let us address the question; given a specific start time, how much time must elapse for the professor to have visited her office at least once with a selected probability? One way of thinking about the question is this: If a student comes to the professor's office, how long must the student wait to be 90% confident (or 99.999% confident) that the professor will show up?

The student could show up at the professor's office and wait. Alternatively, the student could check the office periodically. As the frequency with which the student checks the office approaches infinity, the elapsed time to finding the professor must converge to the period

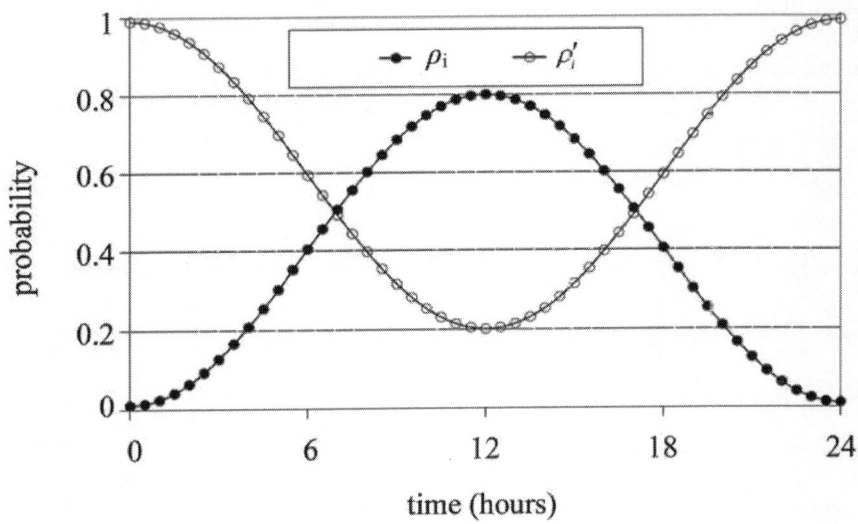

Figure 1. Probability ρ of finding a professor in her office as a function of time together with the probability ρ' that the office is not occupied.

of time the student would have to wait at the office door. Convergence of the time to achieve a given confidence level with increasing sampling frequency is therefore a necessary condition for the validity of any approach that is based on repetitive sampling of the probability function. In the next subsection we show that simple multiplication of the probabilities extracted from the probability function fails in this regard.

3.1. Simple Probability Analysis

We begin our analysis of the professor office-occupancy problem by studying what happens if the student checks the professor's office periodically. The naive approach assumes probabilistic independence, in which case the probability that the office has never been found to be occupied in N samplings P_N^{not} is,

$$P_N^{not} = \prod_{k=1,\cdots,N} \left(1 - \rho_k(t)\right), \tag{11}$$

where $\rho_k(t)$ is the probability of the office being occupied in a specific sampling event k. The probability that the professor has been found in her office is then $\left(P_N = 1 - P_N^{not}\right)$. This approach may appear to give reasonable results for widely separated sampling events $\left(\Delta t = t_k - t_{k-1} \gg 0\right)$, but obviously the assumption that $\rho(t)$ and $\rho(t - \Delta t)$ are probabilistically independent is flawed. Consequently, the method fails as $\Delta t \to 0$. This failure can be seen by considering four cases: 1) The student looks for the professor at exactly noon each day. 2) The student looks for the professor at noon and midnight each day. 3) The student looks for the professor at noon and 2 seconds after noon each day. 4) The student looks for the professor at noon and every 2 seconds thereafter until 12:00:08 pm each day.

Applying Equation (11) to the four cases described above yields the results in Table 1, which shows the day on which the probability of having found the professor exceeds 99.999%. In case 1, after N consecutive days the probability that the student has located the professor is $P_N = 1-(0.2)^N$. This expression converges to 1.0 with increasing N, a physically sensible result. In case 2, after M consecutive days the probability that the student has located the professor is $1-(0.198)^M$. This converges to 1.0, but slightly more rapidly than if the student checks only at noon, again a physically sensible result. By checking at midnight in addition to noon, the student increases the probability of finding the professor, but only by a little because the professor is very unlikely to be found in her office at midnight.

For case 3, since

$$\left(\rho'(t=12:00:00) \approx \rho'(t=12:00:02) \approx 0.2\right)$$ after

N consecutive days, the probability of having located the professor will be $1-(0.2)^{2N}$. This result is completely unreasonable. The probability is converging to 1.0 more rapidly than in case 1, only because the student is making two attempts to find the professor at essentially the same time each day! Case 4 further highlights the failure of this approach. Table 1 shows that the professor is found 10 days earlier simply because the student is looking into the office every two seconds for 8 minutes each day, instead of simply checking once.

Table 1. The day the professor is found in her office with 99.999% probability, based on looking for her at different times of the day under the assumption that these samplings possess probabilistic independence.

	Formula to calculate the probability the professor has been found in her office.	Day the probability of finding the professor in her office reaches 99.999%.
Case 1	$1 - 0.2^N$	11th day
Case 2	$1 - (0.2 \times 0.99)^N$	11th day
Case 3	$1 - (0.2 \times 0.2)^N$	6th day
Case 4	$1 - (0.2 \times \cdots 0.2)^N$	2nd day

The reason for the failure of Equation (11) is that $\rho(t)$ is very strongly correlated to $\rho(t-\Delta t)$ for small Δt owing to the continuity of the time evolving probability function $\rho(t)$. In the next section we describe an approach that accounts for the correlation between sampling events to produce a predicted time to achieve a given confidence level that converges with increasing sampling frequency.

3.2. CONDITIONAL PROBABILITY

During a discrete time-step (Δt), the condition of the professor being in the office will either change or stay the same. In general, the probability that the condition changes may depend on the initial state. We can represent the evolution during a time-step with a tree as depicted in Figure 2.

In the tree diagram, a is the probability that the "state" changes given that the professor is not in her office and b is the probability that the state changes given that she is in her office

($a' = 1-a$ and $b' = 1-b$). We can write the probabilities of being in the office at $t+\Delta t$ as:

$$\rho_{i+1} = X + Y = \rho_i' \cdot a + \rho_i \cdot (1-b) \quad , \tag{12a}$$

$$\rho_{i+1}' = W + Z = \rho_i' \cdot (1-a) + \rho_i \cdot b \quad . \tag{12b}$$

If a and b can be determined from ρ_i and ρ_{i+1} (and ρ_i' and ρ_{i+1}') it is then possible to determine the probability that the professor was never in her office. The probability is given by the probability that the office was not occupied at time $t=0,(\rho_0')$ times the probability that the state did not change on each subsequent time-step $(a'=1-a)$. In other words, given that the office was not occupied at time $t=i$, if the probability that the state did not change between time $t=i$ and time $t=i+1$ is given by $a_{i,i+1}'$, then the probability that the office was never occupied is given by,

$$P^{not} = \rho_0' \prod_{i=0,\cdots,N} a_{i,i+1}' = \rho_0' \prod_{i=0,\cdots,N} \left(1-a_{i,i+1}\right) \quad . \tag{13}$$

In fact, Equations (12a) and (12b) are dependent and we can therefore only solve for a in terms of b. While a definitive relationship between a and b is not obvious, both a and b values must be confined by $0 \le (a,b) \le 1$ because they represent probabilities. By rearranging Equations (12a) and (12b), we obtain the following expressions.

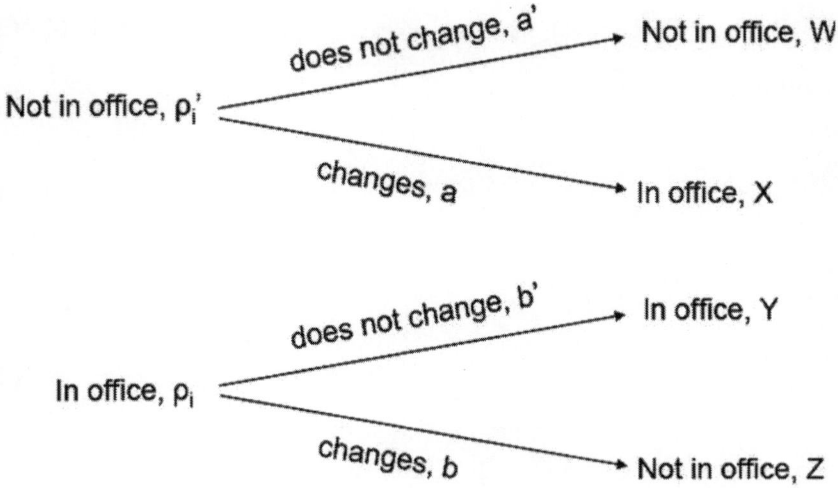

Figure 2. Conditional probability tree diagram where: ρ_i is the probability that the professor is in her office at time $t = i, p_i'$, is the probability that she is not in her office at time $t = i$, and $\rho_i' = 1 - \rho_i$, at any time t_i.

$$a = (\rho_i / \rho_i')b + (\rho_i' - \rho_{i+1}') / \rho_i' \tag{14a}$$

$$b = (\rho_i' / \rho_i)a + (\rho_{i+1}' - \rho_i') / \rho_i . \tag{14b}$$

Using probability function Equation (10) and the relationships of (14a) and (14b), we can find the range of allowed values of a and b at each time-step.

Consider as a specific example, the set of conditions: $\rho_i = 0.6$, $\rho_i' = 0.4$, $\rho_{i+1} = 0.7$, $\rho_{i+1}' = 0.3$. It follows that for this specific set of conditions,

$$b = (4/6)a - 1/6 \quad (15)$$

We haven't identified another relationship between a and b, but we do know that $[0 \le a \le 1]$ and $[0 \le b \le 1]$. The range of allowed values of a and b for this example case is shown with the boldface line segment on the LHS graph of Figure 3.

Based on analyzing the allowed ranges for a and b, the minimum and maximum allowed values of $a(b=0,1)$ and $b(a=0,1)$ using two sampling timesteps, (0.5 hour and 0.1 hour) are shown in Figures 4(a) and 4(b). From the figures we see that $a_{\min}$ and $b_{\min}$ decrease as the time-step gets smaller. This is a physically sensible result: As the time step decreases, the probability that the state changes during the time step also decreases (i.e., the state is very unlikely to change during a very short time interval).

There are many mathematically allowed solutions of Equation (14) for a and b. To obtain the physically meaningful solution, a second condition on a and b is required. One obvious possibility is $a + b = 1$. Using this condition, direct application of Equation (13) produces the same unphysical result as the simple probability approach. The predicted time to achieve a given confidence level does not converge with decreasing time-step. We conclude that this obvious additional condition on a and b is not the physical one.

We note that physically, a and b represent the probabilities of the "occupied" and "unoccupied" states changing during the given time interval. While we do not know the specific values of a and b for any given time interval, it is physically reasonable that as the time step is decreased, the probability of the state changing during any one time step decreases. In the limit of decreasing time step therefore, a and b should take on their smallest allowed values. We therefore used the slowest mathematically allowed flow of probability by using $a+b=\text{minimum}$ where we denote $a=a_{\min}$ and $b=b_{\min}$. The case $(a+b=\text{minimum})$ represents the fewest office occupancy state changes that produce ρ_i' and ρ_{i+1}', from ρ_i and ρ_{i+1}. This is the physically meaningful additional condition on a and b.

In the range $[0<t<12]$, the probability that the office is occupied is increasing, therefore the minimum change is accomplished by setting $b=0$, which produces $a=a_{\min}$. Here, if the office is occupied it remains so, if it is not, there is a small probability of a change in the state $(a_{\min})$. Similarly, for $[12<t<24]$, the probability that the office is occupied is decreasing. The minimum change is accomplished by setting $a=0$, which produces $b=b_{\min}$. In this case, if the office is empty it remains so, but if it is occupied there is a small probability of a change in the state $(b_{\min})$.

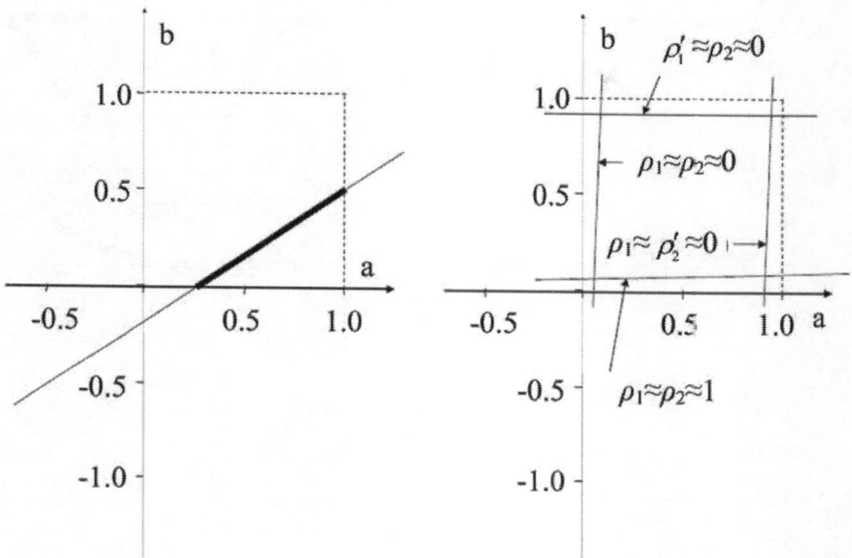

Figure 3. LHS: allowed ranges for a and b for the set of conditions: $\rho_i = 0.6$, $\rho_i' = 0.4$, $\rho_{i+1} = 0.7$, $\rho_{i+1}' = 0.3$. RHS: extrema of allowed values of a and b in general.

One might argue that the state of the system could change during the first time interval (or any subsequent time interval) regardless of how short the time increment is, in which case $a = 1$, not its minimum allowed value. This is true, however, we seek to place a confidence on the office occupancy within some elapsed time. While the rapid state change may happen, to specify a confidence that the state change has happened we must base our analysis on the slowest rate of state change that is consistent with the probability function.

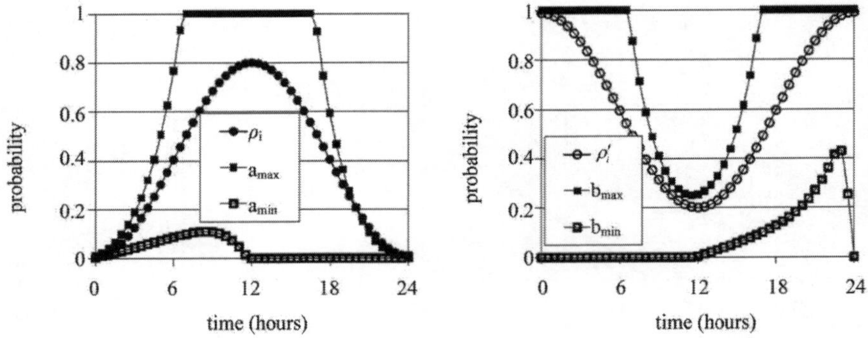

(a)

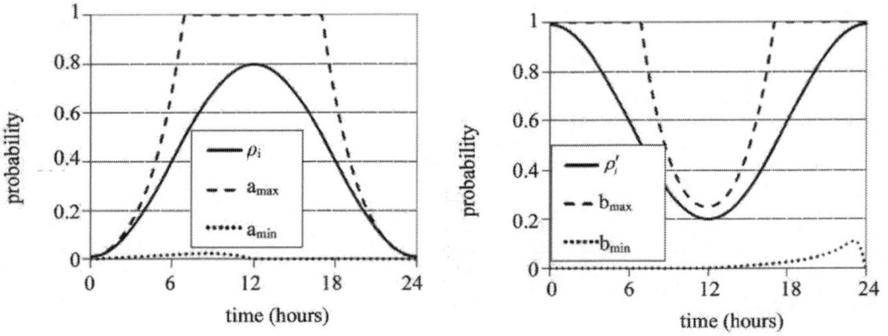

(b)

Figure 4. (a) Allowed range of a and b values for conditional probability for professor office-occupancy problem. Also plotted are the $\rho(t)$, with a_{max}, a_{min} and $\rho'(t)$ with b_{max}, b_{min}. The plots are made for 0.5 hour time intervals. (b) Allowed range of a and b values for conditional probability for professor office-occupancy problem. Also plotted are the $\rho(t)$, with a_{max}, a_{min} and $\rho'(t)$ with b_{max}, b_{min}. The plots are made for 0.1 hour time intervals. The spacing of data points is so small that individual points have been replaced with an interpolating curve for visual clarity. Note that a_{min} and b_{min} are smaller than in Figure 4(a) where a larger time step was used.

Using the above analysis, we can put rigorous upper and lower bounds on the flow of probability. To obtain the upper limit, we use $a = a_{\max}$. In this case the office is occupied the first time we look. To obtain the lower limit we use $a = a_{\min}$, which produces $P^{not}(t)$ as shown in Figure 5. Note that P^{not} drops below 0.1 at 30.5 hours. (Half-hour time steps were used to generate the figure). This is the elapsed time at which we can be 90% confident that the professor has been in her office at least once.

The results (collected in Table 2) show that as the time-step gets smaller, the approaches using simple probability, and conditional probability using $a + b = 1$, predict that the time to 90% occupancy probability decreases to zero with increasing sampling frequency, a physically nonsensical result. Only the conditional probability analysis with the assumption of minimum flow produces a result that converges.

3.3. V alidation by Monte Carlo Simulation

3.3.1. Simulation Methods

To validate our findings, we performed Monte Carlo (MC) simulations based on the time-dependent probability function (10) for both the simple probability and conditional probability approaches. We compared the results to direct application of equations (11—simple probability been in her office at least once (τ_{90}). The boldface result is the physically meaningful one. The other results are based on the flawed assumption that samplings of the probability function are independent events. Monte-Carlo denotes the result of a Monte-Carlo simulation. Direct denotes direct application of the noted equation.

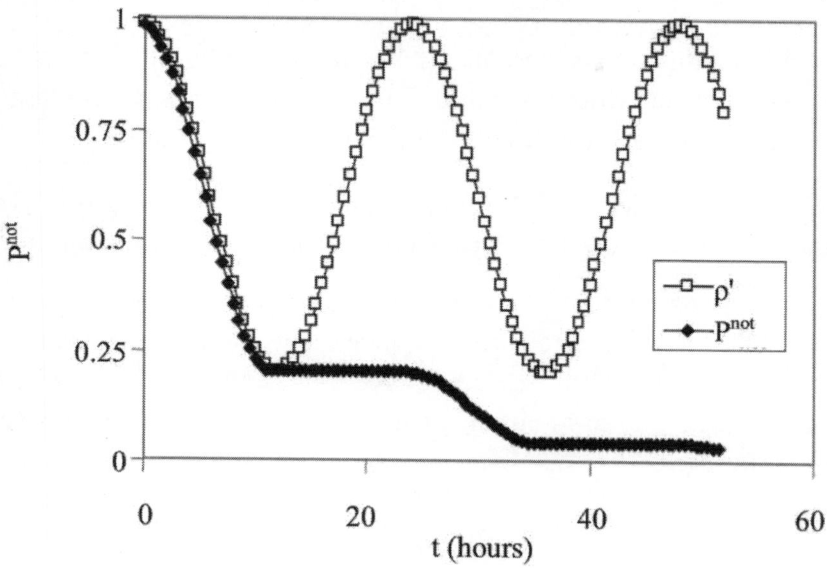

Figure 5. Time dependence of ρ' (the probability that the office is not occupied) and P^{not} (the probabilistic confidence that the office has never been occupied between $t = 0$ and t) based on conditional probability analysis.

Table 2. The elapsed time (hours) at which the professor is 90% likely to have analysis) and (13—conditional probability analysis). For consistency, in all cases we determined the elapsed time when the professor is 90% likely to have been in her office at least once(τ_{90}).

approach →	Simple probability (Equation (11))		Conditional probability (Equation (13))		
time-step (hour) ↓	τ_{90} direct	τ_{90} Monte-Carlo	τ_{90} $a+b=1$ direct	τ_{90} a_{min}, b_{min} direct	τ_{90} a_{min}, b_{min} Monte-Carlo
60 min (1 hr)	9.00	8.00	8.00	31	30.5
30 min (1/2 hr)	6.50	6.00	6.00	31	30.7
15 min (1/4 hr)	5.00	4.75	4.75	31	30.8
10 min (1/6 hr)	4.33	4.17	4.17	31	30.9
1 min (1/60 hr)	1.68	1.65	1.68	31.0	30.9

In the first set of MC simulations, (simple probability) the "1st visit time" is found by starting at time $t=0$ of the first day and advancing in discrete time steps Δt. At each time-step a random number is chosen between 0 and 1. If the random number falls under the probability curve $\rho(t)$ the professor is taken to be in her office and the answer is obtained. If the random number falls on or above the curve, another time-step is taken and so on until we get the 1st visit. We look at all first visits and find the time of day when 90% of the first visits have occurred. We continued to follow the professor's movement beyond the 1st visit to validate whether the simulation procedure reproduces the $\rho(t)$ profile. (This allows us to look at the office-occupancy problem without an artificial day-cycle boundary.) One or multiple consecutive in-office-states constitute a single visit. Most of the visits are in the middle of the day-cycle as expected. This approach reproduces the $\rho(t)$ profile with the professor spending on average 9.72 hours per day in the office as shown in Figure 6.

For comparison, for direct application of Equation (11), the time the professor is found in her office at least once with 90% probability is determined by setting the value of Equation (11) to 0.1 and solving for t. Here k runs from 0 to N and $t=k\cdot\Delta t$. The probability is given by the probability that the professor was not in her office at time $t=0$, (ρ_0') times the probability that she was not in her office at the end of each subsequent time-step (ρ_i'). The product is carried out stepwise, incrementing N by 1 until P_N^{not} drops below 0.1.

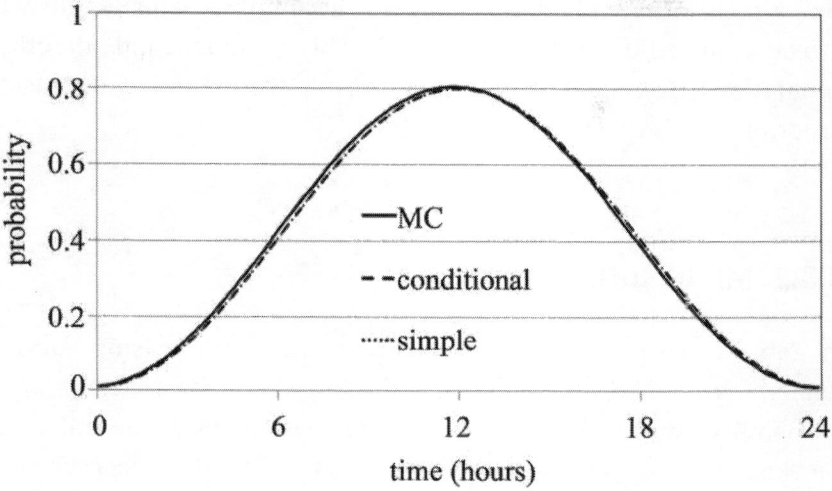

Figure 6. Comparison of $p(t)$ profiles based on simulations with $p(t)$ function of Equation (1): MC = MonteCarlo simulation, conditional = simulation with conditional probability with minimum flow, simple = simulation based on simple multiplication of probabilities. For the MonteCarlo simulations 10,000 day-cycles were used. All of the simulations reproduce the original $p(t)$ function and predict an average of 9.72 hours spent in office per day.

A Monte Carlo simulation based conditional probability with minimum flow is obtained in a very similar manner as the simple probability approach described two paragraphs above. but with one difference; at each timestep we compare the random number to the a_{min} in the range $0 < t < 12$ and b_{min}, for the range $< 24 \; 12 < t$, instead of comparing the random number to $\rho(t)$ as done in the simple probability approach.

Direct application of Equation (13) was accomplished analogously to direct application of Equation (11). This product, and all other simulations described in this manuscript, were carried out using program MATLAB [18].

3.3.2. MC Results

In Table 2 we report the results of our calculation using varying time-steps. The results show that the answer given by the simple probability approach depends on how frequently the office is checked and converges to zero as the time step is decreased, a physically nonsensical result. Simply checking the office with infinite frequency does not guarantee that the professor will arrive immediately!

As shown in Table 2, the simulation based on the condition of minimum flow produces an answer that is independent of time-step and is exactly the same as the analytic analysis shown in Figure 5, and the direct application of Equation (13). The minimum flow conditional probability approach also produces the correct $\rho(t)$ profile as shown in Figure 6 with a total office time of 9.72 hours per day-cycle.

In the approach using conditional probability with minimum flow rate, since $a = a_{\min} = 0$ from noon till midnight, no first visits are found during the final 12 hours of a day-cycle. As a result, some days have no visits at all and the answer, time 'till the first visit, is greater than 24 hours (It is important to note that the professor can arrive at her office within the first 24 hours, or 12 hours or any shorter time interval, but 90% confidence is not achieved until later). This approach achieves our requirement of obtaining a solution to the problem that converges with decreasing time-step.

There is an important feature of quantum mechanics that merits note here: In a quantum system, the timedependent wavefunction (and its corresponding probability density $\rho(t)$) evolves according to the TDSE until the state of the system is observed, whereupon the system collapses into a stationary state. In practice therefore, the process that we draw analogy to, systematically and repeatedly peeking into a professor's office to ascertain its occupancy, would function for that classical case, but not for a quantum mechanical system. In a quantum system, the first such "measurement" would collapse the wavefunction into a stationary state. It is important to draw a distinction, however, between carrying out a measurement of the system and evaluating a probability density function. To analyze the mathematical properties of $\rho(t)$ we may evaluate it at numerous values of t without altering the function, just as integrating the absolute square of a 1D spatial wavefunction

$|\psi(x)|^2$ over the interval $[-\infty \leq x \leq \infty]$ to establish that it is properly normalized does not collapse the system into a single-valued position state, which would be represented by a delta function. For this reason, we may use the analogy to the classical office occupancy case to inform us of how to analyze the time evolving probability function. In the ensuing sections we apply the minimum flow conditional probability analysis to two quantum mechanical systems.

4. ELECTRON ORBITAL-OCCUPANCY

We have applied the methods developed above to the orbital-occupancy problem where $\rho(t) = \sin^2(\beta t/\hbar)$ gives the time dependence of electron occupancy for orbital χ_2 (See Equation (9b)). The oscillation frequency is $\beta/\hbar$. In our calculations we took the values of β and $\hbar$ to be 1. The cycle length is therefore π (The function repeats every time t advances by π). The results of our studies for the electron orbital occupancy problem are collected in Table 3, which shows the elapsed time (fraction of a cycle) at which the electron is 90% likely to have occupied the second orbital at least once. Note that just as in the professor office-occupancy problem, the simple probability approach produces the physically unreasonable prediction that as the sampling rate increases, the time to 90% occupancy probability converges to zero. The simple probability and $a+b=1$ Monte Carlo simulations produce the same unphysical result because they are based on the same flawed assumption of uncorrelated probabilities. The approach of using conditional probability with minimum flow rate converges to 1.25, indicating that at about 125/314th of an oscillation cycle, the electron has reached the second "acceptor" orbital with 90% probability.

While the results presented here employ a basis of two atomic orbitals, the extension to a larger basis is straightforward. Numerical integration of the TDSE for a non-stationary state representing an initially localized electron yields a discrete representation of the time dependent state vector. Upon transformation from the basis of eigenstates of the TISE to the AO basis, the squares of the AO expansion coefficients are the time dependent occupancy probabilities. Application of the conditional probability analysis described above to a selected orbital will give the elapsed time to expected occupancy of that orbital.

While our example analyses the time dependent probability function arising from a non-stationary wavefunction evolving under the influence of a time-independent Hamiltonian, it can, in principle, be applied more generally. The time-evolving probability need not be based on a time-independent Hamiltonian. In the case of a timedependent Hamiltonian, the methods of finding the time-dependent wavefunction would be different, but our method of analyzing the probability function would still apply.

5. PROTON TRANSFER ACROSS AN ASYMMETRIC DOUBLE WELL

As a second application of our conditional probability analysis we consider the exchange of a proton across an asymmetric double-well potential. Here we employ the quantum mechanical model of Hameka and de la Vega [19]. The Hamiltonian is taken to be time-independent and of the form,

$$\hat{H} = H^0 + H'. \tag{16}$$

Here H^0 describes a symmetric double-well potential having the lowest two eigenenergies $E^0 \pm \delta$, where δ represents half the splitting of these lowest two eigenenergies. The fundamental oscillation frequency is $\omega_0 = 2\delta/\hbar$. The perturbation term, $H' = \Delta \frac{x}{d}$, introduces the asymmetry to the potential. Here Δ is the energy difference between the classical minima of the asymmetric double well, d is their spatial separation and x is the displacement coordinate. Based on this model, Hameka and de la Vega [19] define a general asymmetry parameter $\rho = \Delta/2\delta$, and show that the oscillation frequency (ω) as a function of the asymmetry parameter is,

$$\omega = \omega_0 \sqrt{1+\rho^2}. \tag{17}$$

Table 3. The time (a.u.) at which the electron is 90% likely to have occupied the second orbital at least once. The boldface result is the physically meaningful one. The other results are based on the flawed assumption that samplings of the probability function are independent events. The oscillation frequency is $\beta/\hbar$, where $\beta=1$ and $\hbar=1$. The cycle length is π.

approach →	Simple probability		Conditional probability		
time-step (a.u.) ↓	τ_{90} direct	τ_{90} by Monte-Carlo Simulation (10,000 cycles sampled)	τ_{90} $a+b=1$ direct	τ_{90} a_{min}, b_{min} direct	τ_{90} by Monte-Carlo Simulation a_{min}, b_{min} (10,000 cycles sampled)
3.14×10^{-1} ($1/10\ \pi$)	1.57	1.26	1.26	1.26	1.10
3.14×10^{-2} ($1/100\ \pi$)	0.63	0.60	0.60	1.26	1.24
3.14×10^{-3} ($1/1000\ \pi$)	0.28	0.28	0.28	1.25	1.25

A traditional measure of lifetime in quantum systems is, $\tau = 2\pi/\omega$. By this measure, the lifetime decreases with increasing asymmetry (increasing $\rho > 1$). We have posed the question; given that the deeper well A is initially occupied, how long must one wait to be 90% confident that the proton has sampled shallower well B? (As before, we denote this period τ_{90}). A plot of τ_{90} versus the asymmetry parameter is given in Figure 7, where it is compared to the traditional measure of lifetime $\tau = 2\pi/\omega$.

Values of the asymmetry parameter corresponding to some of the proton transfer systems considered by Hameka and de la Vega are marked on the plot for reference. Note that even though the oscillation frequency increases with increasing asymmetry, the duration one must wait for the proton to migrate from one well to the other also increases. This is because, even though the frequency of oscillation increases, the probability of any one well being occupied does not oscillate between 0 and 1, but within a narrower range. In the limit of great asymmetry, probability is pooled almost exclusively in a single well and one must wait essentially forever for the proton to migrate, even though the oscillation frequency is very high. This is consistent with the experimentally known lack of tautomerism in 2-methylnaphthazarin [19], for which the asymmetry parameter is 9.3×10^5.

6. CONCLUSIONS

A probabilistic method is presented to compute a transit time for the transport of a quantum particle from one spatially localized state to another. This approach is based on application of conditional probability to discrete sampling of the time dependent probability function (the absolute square of the time-dependent wavefunction), with the additional condition of minimum flow rate. The method is developed by analyzing a probabilistic classical system and is subsequently applied to two cases of quantum two-state oscillations: electron orbital-occupancy and proton transfer. In this approach, the quantum particle is initially localizing in one spatially localized state (orbital) and the time development of the corresponding non-stationary wavefunction of the time-independent Hamiltonian is followed as the particle travels to a second spatially localized state. There is no definitive time at which the particle can be said to have arrived at the second orbital. Instead, we compute the elapsed time to achieve a set probabilistic confidence level (e.g. 90%, 99.999%) that the particle has reached the second state. Unlike approaches based on the (flawed) assumption that the probability that a given orbital is occupied at time t is independent of the probability that the same orbital is occupied at an earlier time t-Δt, the approach yields an answer that converges with decreasing sampling timestep. Application of the method to asymmetric proton transfer yields results that are both consistent with known experimental evidence of tautomerism and more physiccally relevant than the traditional $\tau = 2\pi/\omega$ measure of life-time. The method of calculation developed here gives a new time-dependent way to analyze and quantify the transport of quantum particles.

ACKNOWLEDGEMENTS

KS thanks C. Rosenthal for numerous valuable discussions as well as and K. Shuford, M. Wander, L. S. Penn and Y. Chen for critical reading of the manuscript. This work was supported by the National Science Foundation grant CHE0449595, including a ROA supplement to support Prof. Kim.

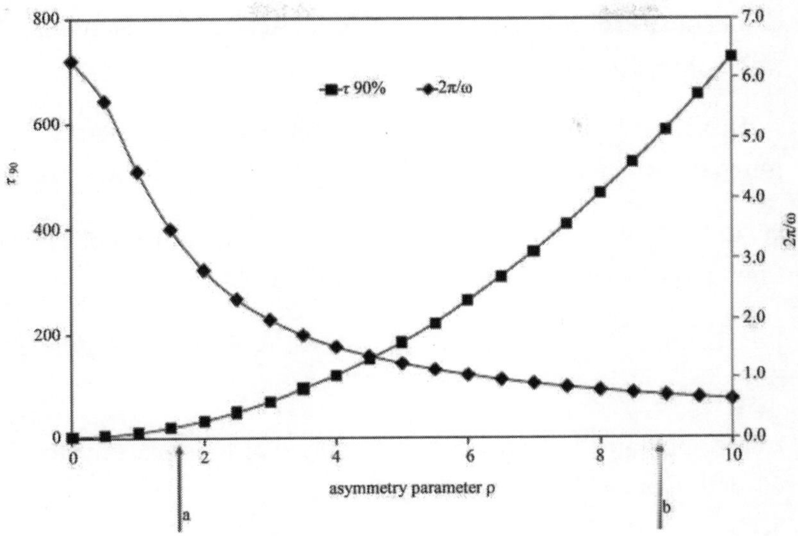

Figure 7. Dependence of τ_{90} on the asymmetry parameter (squares—left vertical axis) in comparison to the traditional measure of quantum lifetime $\tau = 2\pi/\omega$ (diamonds—right vertical axis). Atomic units are used. Arrow a marks the asymmetry parameter of 9-hydroxyphenalen-1-one and b marks the parameter for α-methyl-β-hydroxyacrolein. For comparison, the asymmetry parameter for 2-methylnaphthazarin is 9.3×10^5.

REFERENCES

1. R. Baer and D. Neuhauser, Chemical Physics, Vol. 281, 2002, pp. 353-362.

2. A. B. Pacheco and S. S. Iyengar, Journal of Chemical Physics, Vol. 133, 2010, Article ID: 044105.

3. H. Guo, L. Liu and K.-Q. Lao, Chemical Physics Letters, Vol. 218, 1994, pp. 212-220.

4. C. Joachim, Proceedings of the National Academy of Sciences, Vol. 102, 2005, pp. 8801-8808.

5. A. Nitzan, J. Jortner, J. Wilkie, A. L. Burin and M. A. Ratner, The Journal of Physical Chemistry B, Vol. 104 2000, pp. 5661-5665.

6. W. R. Cook, R. D. Coalson and D. G. Evans, The Journal of Physical Chemistry B, Vol. 113, 2009, pp. 11437- 11447.

7. J.-P. Launay, Chemical Society Reviews, Vol. 30, 2001, pp. 386-397.

8. J. A. Hauge and E. H. Støvneng, Reviews of Modern Physics, Vol. 61, 1989, pp. 917-936.

9. H. G. Winful, Physics Reports, Vol. 436, 2006, pp. 1-69.

10. Y. Aharonov, N. Erez and B. Reznik, Physical Review A, Vol. 65, 2002, Article ID: 052124.

11. J. M. Deutch and F. E. Low, Annals of Physics, Vol. 228, 1993, pp. 184-202.

12. F. E. Low and P. F. Mende, Annals of Physics, Vol. 210, 1991, pp. 380-387.

13. R. S. Dumont and T. L. Marchioro II, Physical Review A, Vol. 47, 1993, pp. 85-97.

14. L. M. Baskin and D. G. Sokolovskii, Russian Physics Journal, Vol. 30, 1987, pp. 204-206.

15. P. Pfeifer, Physical Review Letters, Vol. 70, 1993, pp. 3365-3368.

16. R. J. Gordon and S. A. Rice, Annual Review of Physical Chemistry, Vol. 48, 1997, pp. 601-641.

17. D. Bohm, "Quantum Theory," Prentice Hall, Englewood Cliffs, 1951.

18. Mathworks, MATLAB, 1984-2010.

19. H. F. Hameka and J. R. de la Vega, Journal of the American Chemical Society, Vol. 106, 1984, pp. 7703-7705

Index

M

N

P

Q

R

S

T

U

W